HOMEWORK HELPERS

Geometry

CAROLYN C. WHEATER

CAREER
PRESS
Franklin Lakes NJ

HOMEWORK HELPERS: GEOMETRY
EDITED BY JODI BRANDON
TYPESET BY EILEEN DOW MUNSON
Cover design by Lu Rossman/Digi Dog Design NYC
Printed in the U.S.A. by Book-mart Press

To order this title, please call toll-free 1-800-CAREER-1 (NJ and Canada: 201-848-0310) to order using VISA or MasterCard, or for further information on books from Career Press.

CAREER
PRESS

The Career Press, Inc., 3 Tice Road, PO Box 687,
Franklin Lakes, NJ 07417
www.careerpress.com

Library of Congress Cataloging-in-Publication Data

Wheater, Carolyn C., 1957-
    Homework helpers : Geometry / by Carolyn C. Wheater.
        p.cm. – (Homework helpers)
    Includes index.
    ISBN-13: 978-1-56414-936-7
    ISBN-10: 1-56414-936-6
        1. Geometry—Study and teaching. I. Title. II. Series.

QA461.W44 2007
516—dc22

                                                              2006038184

Dedication

To my father, Pete Catapano, who refused to listen to the people who told him it was foolish to let a girl go to college, and to my father-in-law, Jim Wheater, who saw in me someone who shared his profession, his avocation, and his sense of humor, and welcomed me into the family anyway.

Acknowledgments

I am deeply grateful to Grace Freedson, for setting the project in motion and for inviting me to be part of it. The project could not have been completed without the endless help and patience of my husband, Jim, and my daughter, Laura. My thanks go to Kristen Parkes, Michael Pye, Jeff Piasky, Linda Rienecker, Adam Schwartz, and everyone at Career Press for their patience, enthusiasm, and support, and to The Nightingale-Bamford School, for the sabbatical program that blessed me with six months to devote to my writing.

Contents

Preface

Welcome to *Homework Helpers: Geometry*!

The traditional course in Euclidean Geometry has been the subject of much discussion over the years. People, in equal numbers, have declared it essential or outmoded, but the fact that it persists suggests it has value. Most importantly, what it has, at the moment, is you.

The Geometry You Already Know

You may note that many of the facts of geometry are already familiar to you. Most students learn the important ideas in earlier math courses, but take them just on faith. The purpose of your course this year is less to assimilate information, and more to develop your ability to analyze a situation, make a conjecture, and support your claim with a logical argument. That skill will be valuable in any discipline. You'll develop your ability to make a sound and convincing argument by building a system of postulates and theorems that you can use to prove other claims. Some, such as the Pythagorean Theorem, may be well known to you; others will be new. As you encounter each new statement, your job will be to question: Why is that true? How do you know? Can you convince others?

The Algebra You'll Actually Need

You will also learn to apply the information in the postulates and theorems to solving problems, and you may feel as though you've done that before as well. With your growing repertoire of algebraic skills, you will now be able to solve a wider variety of problems.

In a geometry course, you'll be expected to solve linear equations and simple quadratics. You can expect the quadratics to be factorable, but you can, of course, use the Quadratic Formula, if you wish. You should be able to distribute and FOIL, simplify radicals, and solve proportions. You'll also be expected to remember how to graph a linear equation and find the slope of a line. You'll sometimes encounter a problem that involves two variables, so that you need to solve a system of equations, but, again, those are usually simple. If you need to review your algebra, seek out a good review book (try *Homework Helpers: Algebra!*)

Tips to Begin

Set aside a section of your notebook, or a stack of index cards, to record each of the postulates and theorems you learn. Having them all in one place will help you locate the information you need to make your arguments.

Check with your teacher about the level of detail he or she expects in your work. Some teachers will expect you to point out that something equals itself; others will let you take that for granted. Your teacher (or your main textbook) may use different notations than the ones in this book. Follow your teacher's example.

Finally, say what you mean. If you say that two angles are congruent because they're vertical angles and vertical angles are always congruent, you don't have to worry about whether that was a definition, a postulate, or a theorem. Common postulates and theorems will have names that can save you some writing, but what they say is more important than what they're called.

So say what you mean. Speak up. Make your argument. Support your thinking. And enjoy your geometry class.

Deductive Reasoning

Lesson 1-1: Conditional Statements

One of the principal reasons people cite for studying geometry is that it develops your logical powers, so it only makes sense to begin here with a discussion of logic. Logic is a study of the systems or patterns of reasoning used in reaching a conclusion. It looks at whether you are putting your statements together in a way that can be relied on to give you a sensible outcome.

A simple declarative sentence is either true or false. Its opposite, or negation, is false if the original is true and true if the original is false. In the study of logic, it is common to denote a statement by a single letter, such as **p** or **q**, and its negation as **~p** or **~q**. For example, if **p** is the statement *Toronto is in Canada*, then **~p** is the statement *Toronto is not in Canada*.

The *if...then*, or **conditional statement**, is a statement of the form *if p, then q*. In symbols, the conditional is **p → q**. The *if* clause is called the **hypothesis**, and the statement that follows the word *then* is called the **conclusion**.

You can think of a conditional statement as a promise. *If you get an A on your next test, then I will sing opera.* When you earn the A, you have a right to expect me to sing opera. If I do, I have kept my promise, so my statement is true. If I refuse to sing after you earn an A, I have broken my promise, so the statement is false.

What happens if you don't get an A? My promise hinged on your earning an A. If you don't do your part, there is no way that I can break my promise. Whether I sing or not, you can't say I broke my promise, because the promise depended on your A. That means: The only time that a conditional statement is false is when the hypothesis is true and the conclusion is false.

Because many of the statements we use in building a logical argument are some form of conditional statement, a more extensive study of conditionals is a good idea. With any two basic statements, there are actually two conditionals you can form: $p \to q$ or $q \to p$. If you also use the negation of each of the statements, the number of different combinations increases. Four conditionals from this family of statements are of particular interest. They are used commonly enough that they have names and figure prominently in discussions of logic.

Start with two simple declarative sentences.

  p: A person is elected president of the United States.

  q: A person is at least 35 years old.

The original conditional statement is $p \to q$, or *If a person is elected president of the United States, then that person is at least 35 years old*. The **converse** of this statement is the conditional formed by changing the statements in the hypothesis and conclusion. The converse is $q \to p$, or *If a person is at least 35 years old, then that person is elected president of the United States*.

It is important to read carefully because forming the converse is more than just changing word order. The statement *A person is 35 years old if she is elected president of the United States* is $p \to q$, not $q \to p$. Look for the word *if* and notice which sentence it precedes. The sentence that follows the word *if* is the hypothesis, whether it comes at the beginning or at the end of the sentence.

Form the negation of each of the original sentences.

  $\sim p$: A person is not elected president of the United States.

  $\sim q$: A person is not at least 35 years old.

Two conditionals can be formed from these. The **inverse** of $p \to q$ is the statement $\sim p \to \sim q$, or *If a person is not elected president of the United States, then that person is not at least 35 years of age*. The **contrapositive** is the statement $\sim q \to \sim p$, or *If a person is not at least 35 years old, then a person is not elected president of the United States*.

Example 1

Form the converse, the inverse, and the contrapositive of the conditional statement *If two professional baseball teams are both in the American League, then they can't meet in the World Series.*

Solution: In this statement:

p: Two professional baseball teams are both in the American League.

q: They can't meet in the World Series.

The negations are:

~p: Two professional baseball teams are not both in the American League.

~q: They can meet in the World Series.

The converse, **q → p**, is:

If they can't meet in the World Series, then two professional baseball teams are both in the American League.

The inverse, **~p → ~q**, is:

If two professional baseball teams are not both in the American League, then they can meet in the World Series.

The contrapositive, **~q → ~p**, is:

If they can meet in the World Series, then two professional baseball teams are not both in the American League.

What is it about the converse, the inverse, and the contrapositive of a conditional that earn them names and special attention? The conditional and the contrapositive always have the same truth value. When two statements have exactly the same truth values, they are said to be **equivalent**, which means that, logically, they say the same thing. The contrapositive is equivalent to the original conditional. The converse and the inverse are equivalent to one another, but not to the original conditional or the contrapositive.

Notice that the conditional statement and its converse are not equivalent. One of the errors of logic that people commonly make is to assume that, when a conditional is true, its converse is true as well. If two professional baseball teams are both in the American League, then they can't meet in the World Series, because the World Series is a contest between the American League champions and the National League champions. The converse of this statement is not necessarily true. If two professional baseball teams can't meet in the World Series, it may be because they are both in the American League, but it might also be because they are both in the National League.

Example 2

Determine which of the following statements are equivalent.

a. If the temperature rises above 90 degrees Fahrenheit, then the air conditioning is turned on.

b. If the air conditioning is turned on, then the temperature rises above 90 degrees Fahrenheit.

c. If the air conditioning is not turned on, then the temperature doesn't rise above 90 degrees Fahrenheit.

d. If the temperature doesn't rise above 90 degrees Fahrenheit, then the air conditioning is not turned on.

Solution: Writing the sentences symbolically may make the analysis easier. Assign letters for the basic statements.

p: The temperature rises above 90 degrees Fahrenheit.

q: The air conditioning is turned on.

Translate each sentence into symbolic form.

a. $p \to q$ (conditional)

b. $q \to p$ (converse)

c. $\sim q \to \sim p$ (contrapositive)

d. $\sim p \to \sim q$ (inverse)

The conditional in **a.** is equivalent to the contrapositive in **c.**

The converse in **b.** and the inverse in **d.** are equivalent to each other.

The name **biconditional** is given to the conjunction of a conditional and its converse:

If a basketball player makes a foul shot, then he scores one point, and if he scores one point, then a basketball player makes a foul shot.

The shorter form of the biconditional is, in symbols, **p ↔ q**, and in words **p** *if and only if* **q**, or:

> A basketball player makes a foul shot if and only if he scores one point.

The biconditional is true when **p** and **q** are both true and when **p** and **q** are both false.

Example 3

Use the given **p** and **q** to write the biconditional **p ↔ q**.

p: You play computer games.

q: You have finished your homework.

Solution: The short form of the biconditional connects **p** and **q** with the phrase *if and only if*, so the biconditional is:

> You play computer games if and only if you have finished your homework.

The long form expresses the biconditional as a conditional and its converse, joined by the conjunction *and*. The long form is:

> If you play computer games, then you have finished your homework, and if you have finished your homework, then you play computer games.

Lesson 1-1 Review

1. Write the converse, inverse, and contrapositive of each conditional.
   a. If I finish my history paper tonight, then I will go to the party tomorrow.
   b. If the river reaches flood stage, then residents will be evacuated.

2. Write both the long and the short form of the biconditional **p ↔ q**, given the following statements.

   p: We spend the weekend in Cape Cod.

   q: The weather is warm and sunny.

Lesson 1-2: Valid Reasoning

Because you want to be confident that your logic is sound, it is good to know some of the commonly accepted patterns of argument. You might think of them as rules for the structure of an argument, but they are just patterns that you can be certain are valid.

The most common and most direct pattern of reasoning with conditional statements is called **detachment**. In symbolic form, the argument is $[(p \rightarrow q) \text{ and } p] \rightarrow q$, but an example may make it clearer. Suppose that you know two things:

> If there is a stop sign at the intersection, then you must stop.

> There is a stop sign at the intersection.

The conclusion is obvious: *You must stop.* Knowing that the conditional $p \rightarrow q$ and p are both true allows you to conclude that q is true.

Indirect reasoning is represented symbolically as $[(p \rightarrow q) \text{ and } \sim q] \rightarrow \sim p$, but again an example is helpful. You know:

> If there is a stop sign at the intersection, then you must stop.

> You don't need to stop.

If you concluded that *there is no stop sign at the intersection*, you reasoned correctly. What you are doing is actually reasoning from the contrapositive. Whenever a conditional is true, its contrapositive is also true. The contrapositive of the conditional above is:

> If you don't need to stop, then there is no stop sign at the intersection.

Combine that with the other known statement (*You don't need to stop*), and you can conclude that *there is no stop sign at the intersection*.

Although these methods are powerful in their own way, they don't take you very far from what you started out knowing. The rule that really lets you move into new territory is **syllogism**. Symbolically, syllogism is $[(p \rightarrow q) \text{ and } (q \rightarrow r)] \rightarrow (p \rightarrow r)$. It may remind you of the transitive property of equality. Consider this example:

> If there is a stop sign at the intersection, then you must stop.

> If you must stop, then you will be late for school.

If you see the linkage and follow it through, you can conclude:

If there is a stop sign at the intersection, then you will be late for school.

Repeated application of this rule can take you far:

If there is a stop sign at the intersection, then you must stop.

If you must stop, then you will be late for school.

If you are late for school, then you will have to serve detention.

If you have to serve detention, then you will miss practice.

If you miss practice, then you won't swim in this week's meet.

If you don't swim in this week's meet, then you won't be considered for an athletic scholarship.

Therefore, if there is a stop sign at the intersection, then you won't be considered for an athletic scholarship.

Example 1

What conclusion, if any, can be drawn from each pair of statements?

a. If the oven temperature is too high, then the roast will burn.

The oven temperature is too high.

b. If your hand shakes, then you color outside the lines.

You color outside the lines.

c. If the ambulance turns on its siren, then you must yield the right of way.

You don't have to yield the right of way.

d. If you rent a DVD, then you must return it tomorrow.

If you must return the DVD tomorrow, then you must watch it tonight.

e. If you want to be healthy, then you should eat fruits and vegetables.

If you want to be healthy, then you should exercise regularly.

Solution: Assign labels to the statements.

a. **h:** The oven temperature is too high.

   b: The roast will burn.

   The given statements are of the form **h → b** and **h**; therefore, you can conclude **b:** The roast will burn.

b. **s:** Your hand shakes.

   l: You color outside the lines.

   The given statements are of the form **s → l** and **l**. No conclusion is possible.

c. **a:** The ambulance turns on its siren.

   y: You must yield the right of way.

   The given statements are of the form **a → y** and **~y**; therefore, you can conclude **~a:** The ambulance did not turn on its siren.

d. **r:** You rent a DVD.

   t: You must return it tomorrow.

   w: You must watch it tonight.

   The given statements are of the form **r → t** and **t → w**. The syllogism allows you to conclude that **r → w**, that is, if you rent a DVD, then you must watch it tonight.

e. **h:** You want to be healthy.

   f: You should eat fruits and vegetables.

   e: You should exercise regularly.

   The given statements are of the form **h → f** and **h → e**. No conclusion is possible.

The two most common logical errors are reasoning from the converse and reasoning from the inverse. Generally, people fall into these errors because they assume that the converse and the inverse are true whenever the conditional are true. As you have seen, that is not the case.

Reasoning from the converse occurs when someone knows that a conditional is true and that its conclusion is true and mistakenly concludes that the hypothesis is true. For example, consider these two statements:

If someone exceeds the speed limit, then that person receives a summons.

George received a summons.

It is common to see people try to draw the conclusion that George exceeds the speed limit, but that is not a valid conclusion. It is possible that George received his summons for speeding, but it is also possible that he was cited for jaywalking, or loitering, or failing to signal a turn.

Reasoning from the inverse is a similar fallacy, but assumes the truth of the inverse rather than the converse. Consider these two statements:

If there is a power outage, then the air conditioning won't work.

There is no power outage.

You might be tempted to conclude that the air conditioning will work, but that conclusion is not guaranteed. The air conditioning might be broken, or a circuit breaker might need to be reset. You can't have confidence in the conclusion.

Example 2

Identify the error in each argument.

a. If a cat is a Manx cat, then it has no tail. Elka's cat is not a Manx cat.

Therefore, Elka's cat has a tail.

b. If a dog is a Collie, then the dog has long hair. Margarita's dog has long hair.

Therefore, Margarita owns a Collie.

Solution:

a. Symbolically, the argument is $[(m \rightarrow \sim t) \wedge \sim m] \rightarrow t$. This is an example of reasoning from the inverse. Elka's cat may have a tail, but may have lost its tail due to injury, so the conclusion is not assured.

b. In symbols, this argument is $[(c \rightarrow h) \wedge h] \rightarrow c$. This is an example of reasoning from the converse. Many breeds of dog have long hair. The fact that Margarita's dog has long hair doesn't guarantee that it is a Collie.

Lesson 1-2 Review

Tell whether the argument is valid. If not, explain why.

1. If you don't use sunscreen, then you get sunburn. You don't use sunscreen. Therefore, you get sunburn.

2. If it rains on Friday, then the game will be cancelled. It doesn't rain on Friday. Therefore, the game is not cancelled.

3. If you ski wildly, then you break a leg. You break a leg. Therefore, you ski wildly.

4. If you study a foreign language, then you learn new vocabulary. If you learn new vocabulary, then you can express yourself effectively. If you can express yourself effectively, then you impress your friends. Therefore, if you study a foreign language, then you impress your friends.

5. If I wake early, then I take a long walk. I did not take a long walk. Therefore, I did not wake early.

Lesson 1-3: Direct Proof and Indirect Proof

Throughout a course in geometry, you will see theorems proven and you will be asked to construct proofs as exercises. The two basic types of proof are direct and indirect proof.

Direct proof uses detachment and syllogism to reason from a given hypothesis to a desired conclusion. Each step in the argument must be a true statement, and you must explain how you know that it is true. Justifications may refer to definitions, postulates, or theorems.

Postulates are fundamental statements that are accepted without proof. Theorems are statements that have been proven. It is a good idea to keep a listing of definitions, postulates, and theorems where you can refer to them easily when you are trying to construct a proof.

You don't officially know any postulates or theorems yet, but a few invented postulates will be enough to illustrate the idea of direct proof. Suppose that you know three postulates:

P1: If I go to the library, then I take the bus.

P2: If it rains, then I go to the library.

P3: It rains every Tuesday.

You are asked to prove:

If it is Tuesday, then I take the bus.

Your job, in giving a direct proof, is to start with the hypothesis (*It is Tuesday*), and reason to the conclusion (*I take the bus*). At each step, you must explain how you know that what you say is true. This proof might go this way:

If it is Tuesday, then it rains, because, according to P3, it rains every Tuesday. If it rains, then I go to the library, according to P2. If I go to the library, then I take the bus, by P1. Therefore, if it is Tuesday, then I take the bus.

The proof uses a syllogism that can be represented by a diagram. Each box is a statement or idea, and each arrow is labeled with the postulate that allows you to make the connection.

Figure 1.1

Example 1

Given the postulates:

P1: If I am going to be on television, then I need a haircut.

P2: If I rob a bank, then I get arrested.

P3: If I get arrested, then the local news will cover the story.

P4: If the local news covers the story, then I will be on television.

Prove: If I rob a bank, then I need a haircut.

Solution: Start with the hypothesis (*I rob a bank*), and look for it in the postulates. According to P2, if I rob a bank, then I get arrested. If I get arrested, the local news will cover the story, according to P3. If the local news covers the story, then I will be on television, by P4, and, according to P1, if I am going to be on television, then I need a haircut. Therefore, if I rob a bank, then I need a haircut.

Example 2

Given the postulates:

  P1: If it is a workday, then I eat breakfast.

  P2: If I watch the *Today* show, then it is not Sunday.

  P3: If I don't watch the *Today* show, then I don't eat breakfast.

Prove: If it is a workday, then it is not Sunday.

Solution: Start with the hypothesis (*It is a workday*), and look for it in the postulates. According to P1, if it is a workday, then I eat breakfast. To make the next link, you need a postulate that begins "if I eat breakfast." None of the given postulates in our list have that hypothesis, but P3 does talk about breakfast. Before you give up, take a closer look at P3. Remember that a conditional statement and its contrapositive are equivalent, so you can use them interchangeably in a proof. Write the contrapositive of P3, and see if it will let you continue your proof. The contrapositive says if I eat breakfast, then I watch the *Today* show. That is helpful, so you can continue the proof by saying if I eat breakfast, then I watch the *Today* show, by the contrapositive of P3. If I watch the *Today* show, then it is not Sunday, according to P2. Therefore, if it is a workday, then it is not Sunday.

Indirect proof is based on reasoning from the contrapositive. In indirect proof, you begin by taking the statement that needs to be proven, and thinking about its contrapositive. It can be helpful to write out the contrapositive. You begin from the hypothesis of the contrapositive and try to build a logical chain to its conclusion. If you can prove the contrapositive is true, the original statement will be true as well.

Suppose you have the postulates:

  P1: If a ball falls within the court, then it is in play.

  P2: If a ball is in play, then you strike it before its second bounce.

  P3: If a ball hits the line, then it is within the court.

P4: If you strike the ball before its second bounce, then the point is not over.

You are asked to prove that if the point is over, then the ball did not hit the line.

Start by writing the contrapositive of the statement you are asked to prove. The contrapositive is:

If the ball did hit the line, then the point is not over.

Use your postulates to build a proof of the contrapositive. If the ball did hit the line, then it is within the court, by P3. If the ball is within the court, then it is in play, by P1. If the ball is in play, by P2, you strike it before the second bounce. If you strike the ball before its second bounce, then the point is not over, by P4. Therefore, if the ball did hit the line, then the point is not over.

Because the contrapositive is proven, the original statement is also true. Therefore, if the point is over, then the ball did not hit the line.

Example 3

Given these postulates:

P1: If we are on vacation, then the berries are not picked.

P2: If it is July, then we are on vacation.

P3: If the berries are not picked, then the birds will eat them.

Write an indirect proof that, if the birds don't eat the berries, then it is not July.

Solution: You want to prove that, if the birds don't eat the berries, then it is not July, so begin by writing the contrapositive: If it is July, then the birds eat the berries. From the hypothesis of this statement, reason to its conclusion: If it is July, then we are on vacation, according to P2. If we are on vacation, then the berries are not picked, by P1. If the berries are not picked, then the birds will eat them, according to P3. Therefore, if it is July, then the birds will eat the berries. This proves the contrapositive, and that means the original conditional is also true. Therefore, if the birds don't eat the berries, then it is not July.

Lesson 1-3 Review

1. Prove that if the budget is not passed, then I can't play the lottery. Use these postulates and write a direct proof:

   P1: If lottery tickets are not sold, then I can't play the lottery.

   P2: If the budget is not passed, then the state government shuts down.

   P3: If the lottery offices are closed, then lottery tickets are not sold.

   P4: If the state government shuts down, then the lottery offices are closed.

2. Write an indirect proof of the statement: If we can make s'mores, then there is no drought. Use these postulates:

   P1: If there is a drought, then the risk of forest fires increases.

   P2: If campfires are banned, then we can't make s'mores.

   P3: If the risk of forest fires increases, then campfires will be banned.

Answer Key

Lesson 1-1

1. a. Converse: If I go to the party tomorrow, then I finish my history paper tonight.

      Inverse: If I don't finish my history paper tonight, then I won't go to the party tomorrow.

      Contrapositive: If I don't go to the party tomorrow, then I don't finish my history paper tonight.

   b. Converse: If residents are evacuated, then the river reaches flood stage.

      Inverse: If the river doesn't reach flood stage, then residents won't be evacuated.

      Contrapositive: If residents are not evacuated, then the river did not reach flood stage.

2. Long form: If we spend the weekend in Cape Cod, then the weather is warm and sunny, and if the weather is warm and sunny, then we spend the weekend in Cape Cod.

   Short form: We spend the weekend in Cape Cod if and only if the weather is warm and sunny.

Lesson 1-2

1. Valid.

2. Invalid. Reasoning from the inverse.

3. Invalid. Reasoning from the converse.

4. Valid.

5. Valid.

Lesson 1-3

1. If the budget is not passed then the state government shuts down, by P2. If the state government shuts down, then the lottery offices are closed, according to P4. If the lottery offices are closed, then lottery tickets are not sold, by P3. If lottery tickets are not sold, then I can't play the lottery, according to P1. Therefore, if the budget is not passed, then I can't play the lottery.

2. The contrapositive of *If we can make s'mores, then there is no drought* would be *If there is a drought, then we can't make s'mores*. Begin with *if there is a drought*, and look to the postulates. If there is a drought, then the risk of forest fires increases according to P1. If the risk of forest fires increases, then campfires will be banned, according to P3. If campfires are banned, then we can't make s'mores. This proves that, if there is a drought, then we can't make s'mores. If that is true, then the original conditional is true as well. So if we can make s'mores, then there is no drought.

Lines and Angles

Lesson 2-1: Fundamentals

Before you can begin a study of geometry, you need to have some basic terms to use in defining and describing other objects. Traditionally, geometry begins with three undefined terms: point, line, and plane. A point is a location. Although represented by a dot, a point takes up no space. A point has no width, no length, and no height. Points are labeled by uppercase letters. A plane is a surface with infinite length and infinite width, but no height or thickness. In a drawing, you can only show a portion of a plane. Planes are usually labeled by uppercase letters.

A line has infinite length, but no width and no height. Lines are always straight; they do not curve or bend, and they go on forever. When drawing lines, put arrows on both ends to indicate that the line continues. Lines are labeled by a single lowercase letter (for example, line l) or by two points on the line. If the line shown is labeled by the two points (R and T) shown on it,

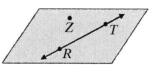

Figure 2.1

it would be written $\overleftrightarrow{RT}$. The line over the top makes it clear that you mean the line.

A ray is a part of a line. In Figure 2.2, ray $\overrightarrow{RT}$ is the portion of line $\overleftrightarrow{RT}$ from point R, through point T, and continuing without end in that direction. Point R is the endpoint (or initial point) of $\overrightarrow{RT}$. Notice that $\overrightarrow{RT}$ and $\overrightarrow{TR}$ are different rays. The order of the points and the direction of the arrow tell you what section of the line forms the ray.

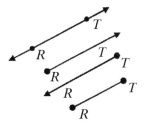

Figure 2.2

A line segment is a portion of a line between two endpoints. A segment is named by its endpoints, with a segment symbol on top. In Figure 2.2, $\overleftrightarrow{RT}$ is a line, but $\overline{RT}$ is a segment of that line, between R and T. Because it has a beginning and an end, a line segment has a measurable length. Two segments that have the same length are congruent.

Lesson 2-2: Rulers and Length

When you use a ruler to measure the length of an object, you usually line up the edge, or zero point, of the ruler with one end of the object and read the number at the other end as the measurement. But if it were impossible or inconvenient to line up with the end of the ruler, you could place the end of the object at any point on the ruler, note the numbers at the ends, and subtract.

Figure 2.3

The **ruler postulate** tells you that you can take any line and put a numbering system on it so that it becomes a ruler. The length of a segment can be found by subtracting the numbers at each end of the segment, called **coordinates**. Because the length is the same whether you measure left to right or right to left, the length is the **absolute value** of the number you get when you subtract. With the segment symbol on top, $\overline{AB}$ means the line segment with endpoints A and B. AB, without a marking on top, means the length of $\overline{AB}$ or distance from A to B. $AB = |b - a|$ where a and b are coordinates or ruler numbers.

Figure 2.4

Example 1

Find the length of $\overline{XY}$.

Figure 2.5

Solution: The coordinate of point X is 7, and the coordinate of point Y is –5. The length of $\overline{XY}$ is $XY = |x - y| = |7 - (-5)| = |12| = 12$. Notice that whether you subtract $7 - (-5) = 12$ or $-5 - 7 = -12$, once you take the absolute value it comes out exactly the same.

Example 2

If $PQ = 4.8$ and the coordinate $q = 1.7$, find p.

Solution: You know that $PQ = |q - p| = 4.8$ and that $q = 1.7$. Technically, you have two possibilities. $|q - p| = 4.8$ could mean that $q - p = 4.8$ or that $q - p = -4.8$, because 4.8 and -4.8 both have absolute value of 4.8. If $q - p = 4.8$ and $q = 1.7$, then $1.7 - p = 4.8$, so $-p = 4.8 - 1.7 = 3.1$ and $p = -3.1$. If $q - p = -4.8$, $1.7 - p = -4.8$, so $-p = -4.8 - 1.7 = -6.5$, and $p = 6.5$. Unless you have a diagram or any other information to help you choose, you should give both answers as possible coordinates for P.

The **segment addition** postulate says that B is between A and C, if and only if $AB + BC = AC$. Many people replace this postulate with a simpler and more general statement of what the postulate says: The length of the whole segment is equal to the sum of the lengths of its parts. Whether you say the whole equals the sum of its parts or you say segment addition, note that $AC = AB + BC$ only if B is between A and C.

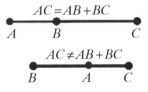

Often in a proof, you need to repeat a sequence of steps you have seen in many proofs.

Figure 2.6

Sequences such as that are highlighted here as **subroutines**. The principal steps are listed, but without reasons, which may vary from proof to proof. You may need some intermediate steps as well, but remember the basic pattern as a technique that may be useful.

If you have information about segments that are the same length, but those segments are larger or smaller than the ones you need, adding or subtracting can solve that problem.

Add Segments

$AB = CD$ $BC = BC$	Reflexive	
$AB + BC = BC + CD$	Addition Property of Equality	
$AC = AB + BC$ $BD = BC + CD$	Segment Addition *(Renaming)*	
$AC = BD$	Substitution	

Figure 2.7

Subtract Segments

$AC = BD$	
$AC = AB + BC$	Segment Addition
$BD = BC + CD$	Segment Addition
$AB + BC = BC + CD$	Substitution
$BC = BC$	Reflexive
$AB = CD$	Subtraction Property of Equality

Figure 2.8

Example 3

Given: $PQ = ST$, $QR = RS$

Prove: $PR = RT$

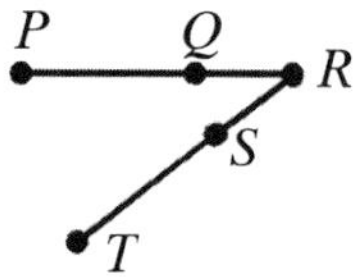

Figure 2.9

Solution: You can start with the first pair of given equal lengths, and add an equal amount to each one to get the bigger piece you want. That is the Add Segments subroutine, but with a little variation. This time, instead of adding the same segment to both, you will add QR to one side and RS to the other. It will be okay, because you are given that $QR = RS$.

Statement	Reason
1. $PQ = ST$	1. Given
2. $QR = RS$	2. Given
3. $PQ + QR = RS + ST$ (Add step #1 + step #2)	3. Addition Property of Equality *(Equals added to equals give equals)*
4. $PR = PQ + QR$	4. Segment Addition *(Renaming: the two segments form one bigger segment)*
5. $RT = RS + ST$	5. Segment Addition *(Renaming: the two segments form one bigger segment)*
6. $PR = RT$	6. Substitution *(Replacing each side of #3 with its new name)*

Figure 2.10

Lesson 2-2 Review

1. If the coordinate of point *P* is −15 and the coordinate of point *Q* is 32, find the length of segment $\overline{PQ}$.

2. If *RS* = 19.5 and the coordinate of *S* is 2.3, find all the possible coordinates for point *R*.

3. Given Figure 2.11 with *AB* = *YZ* and *AW* = *CZ*, prove *BW* = *CY*.

   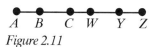

   Figure 2.11

4. If you completed Exercise 3, you now know that in the figure, *BW* = *CY*. Prove *BC* = *WY*.

Lesson 2-3: Protractors and Angle Measure

An angle is a figure formed by two rays with a common endpoint. The common endpoint is the vertex of the angle and the rays are its sides. In the figure, $\overrightarrow{XY}$ and $\overrightarrow{XZ}$ are the sides of angle and point *X* is the vertex. An angle can be named by the letter of its vertex, if there is no possibility of confusion (for example, ∠*X* in the Figure 2.12), or by three letters that outline the angle, with the vertex letter in the middle (∠*YXZ* or ∠*ZXY*).

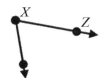

Figure 2.12

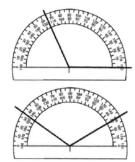

Figure 2.13

When you write ∠*ABC*, you designate an angle. To indicate the measure of an angle, write m∠*ABC*. Two angles that have the same measurement are congruent angles. What determines the measure of an angle is the amount of rotation from one ray (side) to the other, not the lengths of the sides.

A protractor is a device for measuring that rotation. A protractor is made from a semicircle marked with numbers from 0 to 180. Once again, you generally line up the zero mark with one side of the angle and read the measurement where the other side falls, but the protractor postulate tells you that you could line up one side of the angle with any marking on the protractor, note where the other side falls, and subtract. Most protractors have two scales, one running clockwise and one counterclockwise. Be certain you're reading the same scale for both the starting and ending rays.

Example 1

Find the measures of ∠BOE, ∠COF, and ∠AOD.

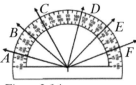

Figure 2.14

Solution: Let's agree to use the top scale, the one that has its zero at the left side.

$\overrightarrow{OB}$ falls at the 40 and $\overrightarrow{OE}$ falls at 135,

so m∠BOE = |135 – 40| = 95. $\overrightarrow{OC}$ falls at 60 and $\overrightarrow{OF}$ at 160, so m∠COF = |60 – 160| = 100.

$\overrightarrow{OA}$ falls at 15 and $\overrightarrow{OD}$ falls at 105, so m∠AOD = |15 – 105| = 90.

Angles are classified according to their size.

Acute	< 90°	Acute means sharp. Acute angles have a sharp point.	
Right	= 90°	Think upright. In a right angle, one side is on the ground and one is upright.	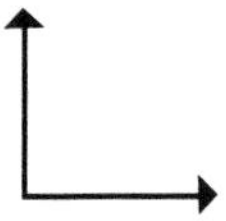
Obtuse	> 90° but < 180°	Obtuse means thick. Obtuse angles look thick.	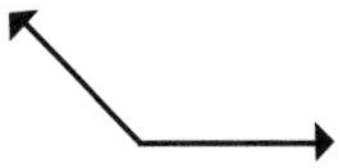
Straight	= 180°	The name says it. Straight angles look like straight lines.	

Figure 2.15

Example 2

If ∠PQR is a right angle and one of its sides falls at 178° on the protractor, what is the coordinate of its other side?

Solution: If ∠PQR is a right angle, its measure is 90°. One of its sides, say $\overrightarrow{QP}$, has a coordinate of 178°. So m∠PQR = 90° = |178° – r|.

Technically, the equation has two solutions.

One is $90° = 178° - r \Rightarrow -88° = -r \Rightarrow r = 88°$.

The other is $-90° = 178° - r \Rightarrow -268° = -r \Rightarrow r = 268°$. Because most protractors only go up to 180°, you should take the first solution. The other side falls at 88° on the protractor.

If two angles have the same vertex and share a side, but do not overlap one another, they are **adjacent angles**. In the figure, $\angle APB$ and $\angle BPC$ are adjacent angles. They have the same vertex, P, and share side $\overrightarrow{PB}$, but they do not overlap. $\angle APB$ and $\angle APC$ are not adjacent angles because $\angle APB$ overlaps $\angle APC$.

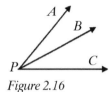

Figure 2.16

Subroutines

Add Angles

$m\angle MON = m\angle QOP$		
$m\angle NOP = m\angle NOP$	Reflexive	
$m\angle MON + m\angle NOP = $ $m\angle QOP + m\angle NOP$	Addition Property of Equality	
$m\angle MON + m\angle NOP = m\angle MOP$ $m\angle QOP + m\angle NOP = m\angle QON$	Angle Addition	
$m\angle MOP = m\angle QON$	Substitution	

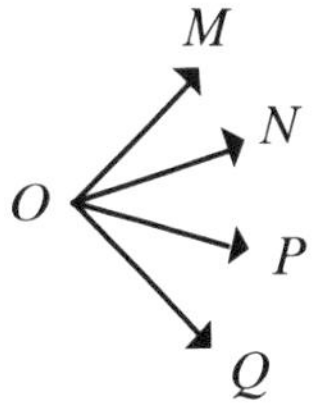

Figure 2.17

Subtract Angles

$m\angle MOP = m\angle NOQ$		
$m\angle MOP + m\angle MON = m\angle NOP$ $m\angle NOQ + m\angle NOP = m\angle POQ$	Angle Addition	
$m\angle MON + m\angle NOP = $ $m\angle NOP + m\angle POQ$	Substitution	
$m\angle NOP = m\angle NOP$	Reflexive	
$m\angle MON = m\angle POQ$	Subtraction Property of Equality	

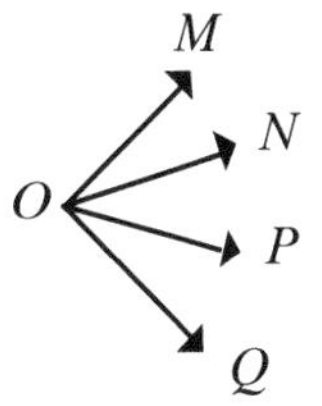

Figure 2.18

1. Find m∠BEC in Figure 2.19 if m∠DEC = 73° and m∠BED = 88°.

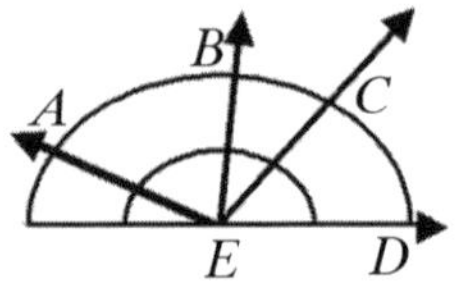

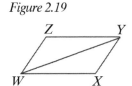

Figure 2.19

2. If the coordinate of $\overrightarrow{EA}$ in Figure 2.19 is 171 and the coordinate of $\overrightarrow{EC}$ is 73, find m∠AEC.

3. **Given:** In Figure 2.20, m∠ZYW = m∠XWY, m∠ZWY = m∠XYW
   Prove: m∠ZWX = m∠XYZ

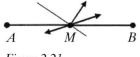

Figure 2.20

Lesson 2-4: Bisecting Segments and Angles

The **midpoint** of a segment is a point on the segment that divides it into two equal parts. The **bisector** of a segment is any line (or ray or segment) that passes through that midpoint. A line segment has only one midpoint, but it can have many bisectors. Segment $\overline{AB}$ has midpoint M. AM = MB and each of the segments $\overline{AM}$ and $\overline{MB}$ is half the length of $\overline{AB}$. Figure 2.21 shows three different bisectors of $\overline{AB}$, all passing through M.

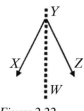

Figure 2.21

Angles don't have midpoints, but any line, ray, or segment that divides the angle into two equal angles is an **angle bisector**. An angle has only one bisector. The two equal angles have the same vertex as the big angle. $\overline{YW}$ bisects ∠XYZ, cutting it into ∠XYW and ∠WYZ. ∠XYW = ∠WYZ and each of these smaller angles is half of ∠XYZ.

Figure 2.22

Use Figure 2.23 with examples 1, 2, and 3.

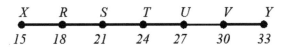

Figure 2.23

Example 1

Which point on segment $\overline{XY}$ is the midpoint of $\overline{RV}$?

Solution: By subtracting the coordinates of R and V, $|30 - 18|$, you find $\overline{RV}$ is 12 units long. Its midpoint should create two segments of length 6. Six units from R is the midpoint, T.

Example 2

What is the coordinate of the midpoint of $\overline{XU}$?

Solution: Subtracting the coordinates tells you the length of $\overline{XU}$ is $|27 - 15| = 12$, so its midpoint will be 6 units from X, at a coordinate of 21.

Example 3

T is the midpoint of a segment that has Y as one of its endpoints. What is the other endpoint?

Solution: The length of $\overline{TY}$ is $|33 - 24| = 9$ so the other endpoint is 9 units below T. $24 - 9 = 15$, so the other endpoint is X.

Example 4

Which ray is the bisector of $\angle AOD$?

Solution: $\angle AOD$ has a measure of $|105° - 15°| = 90°$, so the bisector divides it into two angles, each measuring 45°. Because $15° + 45° = 60°$, the ray that falls at 60°, $\overrightarrow{OC}$, is the bisector of $\angle AOD$.

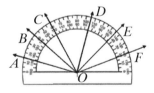

Figure 2.24

Bisectors divide segments or angles into two equal parts, each of which is half of the original. If two segments have equal length, and each of them is bisected, the small pieces are all the same size. If you start with two equal angles, and you bisect each one, you divide them into congruent angles. Each of those smaller congruent angles is half of the original and all the small angles equal one another.

Subroutines

Bisect Segments

$\overline{EF}$ bisects $\overline{AC}$ $\overline{EF}$ bisects $\overline{BD}$		
$AE = \frac{1}{2}AC$ $BF = \frac{1}{2}BD$	The bisector divides a segment into two segments each of which is half as long as the original.	
$AC = BD$		
$\frac{1}{2}AC = \frac{1}{2}BD$	Multiplication Property of Equality	
$AE = BF$	Substitution	

Figure 2.25

Bisect Angles

$\overline{BC}$ bisects $\angle SBI$ $\overline{IE}$ bisects $\angle SIB$		
m$\angle CBI = \frac{1}{2}$ m$\angle SBI$ m$\angle EIB = \frac{1}{2}$ m$\angle SIB$	An angle bisector creates two congruent angles, each half as large as the original.	
m$\angle SBI =$ m$\angle SIB$		
$\frac{1}{2}$ m$\angle SBI = \frac{1}{2}$ m$\angle SIB$	Multiplication Property of Equality	
m$\angle CBI =$ m$\angle EIB$	Substitution	

Figure 2.26

Lesson 2-4 Review

1. $\overleftrightarrow{XB}$ bisects $\overline{WV}$. If $\overline{WY}$ measures 7 units, find the length of $\overline{WV}$.

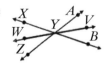

Figure 2.27

2. If $\angle XYZ$ measures 52° and $\overrightarrow{YW}$ bisects $\angle XYZ$, find the measure of $\angle XYW$.

Lesson 2-5: Complementary and Supplementary Angles

Two angles are **complementary** if the total of their measures is 90°. If the sum of the measures of two angles is 180°, the angles are **supplementary**. To remember which word goes with which total, remember that they are in both alphabetical and numerical order: Complementary comes before supplementary alphabetically, and 90 comes before 180 numerically. The words *complementary* and *supplementary* apply only to pairs of angles. You may have three angles that add to 180°, but they are not called supplementary. Complementary angles and supplementary angles don't have to be adjacent angles. If two angles are complementary, each is the *complement* of the other. Two angles that are supplementary to one another are *supplements.*

Example 1

Find the complement of an angle of 23°.

Solution: An angle and its complement add to 90°, so to find the complement, subtract 23° from 90°. 90° − 23° = 67°, so the complement of an angle of 23° is an angle of 67°.

Example 2

The measure of ∠A is x + 10 and the measure of ∠B is 3x + 50. If ∠A and ∠B are supplementary, find x.

Solution: Supplementary angles add to 180, so:

$$(x+10)+(3x+50)=180$$
$$x+3x+10+50=180$$
$$4x+60=180$$
$$4x=120$$
$$x=30$$

Lesson 2-5 Review

1. Find the supplement of an angle of 15°.

2. Find the complement of an angle of 41°.

3. If the measure of $\angle P$ is $5x + 3$ and the measure of $\angle Q$ is $2x + 3$, and $\angle P$ and $\angle Q$ are complementary, find the measure of $\angle P$.

4. $\angle ABC$ and $\angle XYZ$ are supplementary. $\overrightarrow{BD}$ bisects $\angle ABC$ and $\overrightarrow{YW}$ bisects $\angle XYZ$. If $\angle XYZ$ measures $44°$, find the measure of $\angle ABD$.

Lesson 2-6: Linear Pairs and Vertical Angles

To form a **linear pair**, two angles must be adjacent, which means that they have the same vertex and share a side. In order for the adjacent angles to be a linear pair, the exterior sides must form a line. The two angles in a linear pair together form a straight angle, which measures $180°$.

> **Linear pairs are always supplementary.**

Example 1

If $\angle 1 = 15°$, find the measure of $\angle 2$.

Figure 2.28

Solution: $\angle 1$ and $\angle 2$ form a linear pair, so they are supplementary. Subtract $180° - 15° = 165°$. The measure of $\angle 2$ is $165°$.

Example 2

$\angle X$ and $\angle Y$ form a linear pair. If m$\angle X = 15x - 12$, and m$\angle Y = 3x + 3$, find the measure of each angle.

Solution: Because $\angle X$ and $\angle Y$ form a linear pair, they are supplementary. Therefore:

$$\text{m}\angle X + \text{m}\angle Y = 180$$
$$(15x - 12) + (3x + 3) = 180$$
$$18x - 9 = 180$$
$$18x = 189$$
$$x = 10.5$$

m$\angle X = 15x - 12 = 15(10.5) - 12 = 145.5°$ and

m$\angle Y = 3x + 3 = 3(10.5) + 3 = 34.5°$

When two lines intersect, four angles are formed. The angles across the × from one another (as ∠*1* and ∠*3* are in Figure 2.29) are called **vertical angles**. They have a common vertex, and their sides form opposite rays. ∠*2* and ∠*4* are another pair of vertical angles in the figure.

Figure 2.29

> **Vertical angles are always congruent.**

Vertical angles are only formed by straight lines intersecting, so don't be fooled by other x-shaped arrangements. In Figure 2.30, ∠*FES* and ∠*AEL* are not vertical angles, because $\overrightarrow{ES}$ and $\overrightarrow{EA}$ don't form a line.

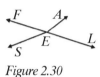
Figure 2.30

Use Figure 2.31 with examples 3 and 4.

Example 3

If m∠*TVI* = 30°, find the measure of ∠*LVE*.

Solution: ∠*TVI* and ∠*LVE* are vertical angles, so m∠*LVE* = 30°.

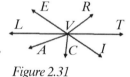
Figure 2.31

Example 4

If $\overrightarrow{VT}$ bisects ∠*RVI*, and m∠*LVE* = 30°, find the measure of ∠*EVR*.

Solution: Because $\overrightarrow{VT}$ bisects ∠*RVI*, m∠*RVT* = m∠*TVI* and because they are vertical angles, m∠*TVI* = m∠*LVE*. That means ∠*RVT*, ∠*TVI*, and ∠*LVE* are all 30° angles. Because they form a straight angle,

$$m\angle LVE + m\angle EVR + m\angle RVT = 180°$$
$$30° + m\angle EVR + 30° = 180°$$
$$m\angle EVR = 120°$$

Lesson 2-6 Review

1. True or False: In Figure 2.31, m∠*AVI* = m∠*EVR*.

2. True or False: In Figure 2.31, ∠*TVI* and ∠*EVT* are supplementary.

3. True or False: The two angles in a linear pair can be complementary.

4. True or False: A pair of vertical angles can be complementary.

5. In Figure 2.32, ∠3 and ∠4 are supplementary, and m∠1 = 155°. Find the measures of ∠2, ∠3, ∠4, ∠5, and ∠6.

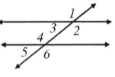

Figure 2.32

Lesson 2-7: Perpendicular Lines

Two lines (or rays or segments) that meet to form a right angle are **perpendicular**. The symbol for "is perpendicular to" is ⊥. If line *l* is per-

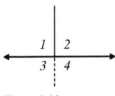

Figure 2.33

pendicular to line *m*, you write *l* ⊥ *m*. Perpendiculars form four right angles, although you may have to extend a segment or ray to see all four of them. You can be certain that, if one of the angles is a right angle, then all four of them will be right angles. You can prove it by using what you know about linear pairs and vertical angles.

Example 1

Determine which line segments in Figure 2.34 are perpendicular if m∠GJK = 65°, m∠JCB = 35°, m∠HJI = 25°, and m∠KCD = 55°.

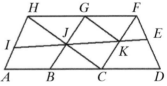

Figure 2.34

Solution: To find perpendicular lines, you need to look for 90° angles. You will need to use what you know about angle relationships to figure out the measures of other angles.

Because you know linear pairs are supplementary, you can determine that m∠BCK = 180° − 55° = 125°.

Because m∠BCK = m∠JCB + m∠JCK, you can calculate that 125° = 35° + m∠JCK, so m∠JCK = 90°.

Because ∠JCK is a right angle, $\overline{JC} \perp \overline{CK}$ or $\overline{HC} \perp \overline{CF}$. In addition, because vertical angles are equal, m∠KJC = m∠HJI = 25°. Then m∠GJC = m∠GJK + m∠KJC = 65° + 25° = 90°. ∠GJC is a right angle, so $\overline{GJ} \perp \overline{JC}$ or $\overline{GB} \perp \overline{HC}$.

Lesson 2-7 Review

1. If $\overline{DO} \perp \overline{ON}$, and $\overline{OE}$ bisects $\angle DON$, what is the measure of measure of $\angle DOE$?

2. If $\overline{DE} \perp \overline{EN}$, $m\angle NEO = 6x - 15$ and $m\angle DEO = 50 - x$, find x.

Figure 2.35

Lesson 2-8: Lines and Angles on the Coordinate Plane

When you work with points in the plane, the Cartesian coordinate system that you learned in algebra lets you assign an ordered pair of coordinates to each point. To find the distance between two points that have the same y-coordinate, say (5, 3) and (9, 3), you only need to subtract the x-coordinates. Likewise, to find the distance between two points on the same vertical line, you simply subtract the y-coordinates. The points $(-4, 2)$ and $(-4, 7)$ are both on the vertical line $x = -4$, so you can subtract the y-coordinates to find that the distance is $7 - 2 = 5$.

When the endpoints of the segment you are trying to measure don't fall on the same vertical line or horizontal line, calculating the distance requires a formula that uses the coordinates of each point. To find that formula, go back to the Pythagorean Theorem. It tells you that, in a right triangle with legs of length a and b and hypotenuse c, $a^2 + b^2 = c^2$. Imagine that your two points are the endpoints of the hypotenuse, and draw the legs as horizontal and vertical line segments. The length of the horizontal segment can be found by subtracting x-coordinates, and the length of the vertical segment by subtracting y-coordinates. That gives you a and b, and you

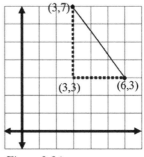

Figure 2.36

can use $a^2 + b^2 = c^2$ to find the value of c. In this example, $a = 6 - 3 = 3$ and $b = 7 - 3 = 4$, so $3^2 + 4^2 = 9 + 16 = 25 = c^2$, and taking the square root tells you that $c = 5$.

In general terms, if the two points are (x_1, y_1) and (x_2, y_2), then $a = x_2 - x_1$ and $b = y_2 - y_1$, so $a^2 + b^2 = c^2$ becomes $(x_2 - x_1)^2 + (y_2 - y_1)^2 = c^2$. Take the square root of both sides, and replace c with d (for distance) to get the **distance formula:**

$$d = \sqrt{(x_2 - x_1)^2 + (y_2 - y_1)^2}$$

Example 1

Find the length of $\overline{PQ}$ if P is the point $(-14, 27)$ and Q is the point $(10, 17)$.

Solution: Use the distance formula. You can call either point (x_1, y_1) and the other (x_2, y_2). Just make your decision and be consistent about it.

$$d = \sqrt{(x_2 - x_1)^2 + (y_2 - y_1)^2}$$
$$d = \sqrt{(-14 - 10)^2 + (27 - 17)^2}$$
$$d = \sqrt{(-24)^2 + (10)^2}$$
$$d = \sqrt{576 + 100} = \sqrt{676}$$
$$d = 26$$

Example 2

You want to draw a segment from the point $(3, -2)$ to a point whose x-coordinate is -3, and you want the segment to have a length of 10. What should be the y-coordinate of the second point?

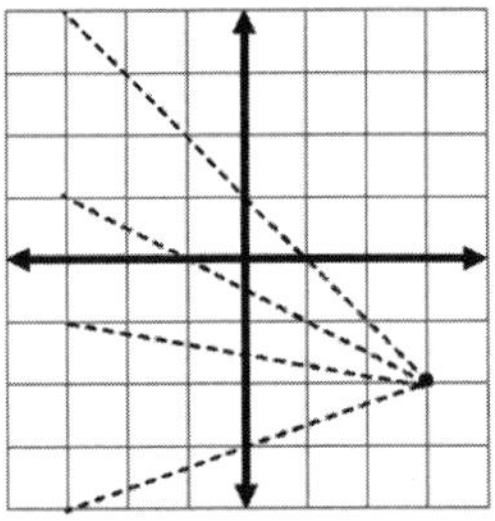

Solution: If you look at some possibilities for the line segment you want to draw, you see that it is going to form the hypotenuse of the right triangle. You know the coordinates of one point, $(3, -2)$, and the x-coordinate of the other, so call that point $(-3, y_2)$. Use the distance formula, and remember that you know the distance is going to be 10.

Figure 2.37

$$d = \sqrt{(x_2 - x_1)^2 + (y_2 - y_1)^2}$$
$$10 = \sqrt{(-3 - 3)^2 + (y_2 - {}^-2)^2}$$
$$10 = \sqrt{(-6)^2 + (y_2 + 2)^2}$$
$$10 = \sqrt{36 + y_2{}^2 + 4y_2 + 4}$$
$$10 = \sqrt{y_2{}^2 + 4y_2 + 40}$$

Square both sides of the equation to get rid of the radical, and solve the quadratic equation.

$$10^2 = \sqrt{y_2^2 + 4y_2 + 40}^{\,2}$$
$$100 = y_2^2 + 4y_2 + 40$$
$$0 = y_2^2 + 4y_2 - 60$$
$$0 = (y_2 + 10)(y_2 - 6)$$
$$y_2 + 10 = 0 \Rightarrow y_2 = -10$$
$$y_2 - 6 = 0 \Rightarrow y_2 = 6$$

You can draw your line segment either to the point $(-3, -10)$ or to the point $(-3, 6)$.

Think about a right triangle, and imagine that you want to find the midpoint of the hypotenuse. The midpoint will have the same x-coordinate as the midpoint of one leg of the triangle and the same y-coordinates as the midpoint of the other leg. So the midpoint, M, can be found by averaging the x-coordinates and averaging the y-coordinates, which we call the **Midpoint Formula**.

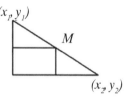

Figure 2.38

$$M = \left(\frac{x_1 + x_2}{2}, \frac{y_1 + y_2}{2} \right)$$

Example 3

Find the midpoint of the segment whose endpoints are $(6, -5)$ and $(-8, 17)$.

Solution:

$$M = \left(\frac{x_1 + x_2}{2}, \frac{y_1 + y_2}{2} \right) = \left(\frac{6 + (-8)}{2}, \frac{(-5) + 17}{2} \right) = \left(\frac{(-2)}{2}, \frac{12}{2} \right) = (-1, 6)$$

The midpoint of the segment is $(-1, 6)$.

Example 4

If $\overline{RT}$ has endpoints R (4, 7) and T (−2, t) and the midpoint of $\overline{RT}$ is (1, 2), find t.

Solution: $M = \left(\dfrac{4+(-2)}{2}, \dfrac{7+t}{2} \right) = \left(1, \dfrac{7+t}{2} \right) = (1, 2)$ so

$\dfrac{7+t}{2} = 2 \Rightarrow 7+t = 4 \Rightarrow t = -3$.

It can be difficult to talk about angles in the coordinate plane without moving into a branch of mathematics known as trigonometry, and that study is for later in the course. You can make a statement, however, about right angles or, more accurately, about perpendicular lines, which form right angles. In algebra, you learned to find the slope of a line. Slope was defined as the ratio of rise to run, or the change in y over the change in x. The formula for the slope of the line through the points (x_1, y_1) and (x_2, y_2) is $m = \dfrac{y_2 - y_1}{x_2 - x_1}$.

The slopes of perpendicular lines are negative reciprocals.

That is, the slopes multiply together to produce −1. If the slope of one line is 5, the slope of a line perpendicular to it will be $-\frac{1}{5}$. If the original line has a slope of $-\frac{3}{7}$, any line perpendicular to it will have a slope of $\frac{7}{3}$.

Example 5

If line l passes through the points (−4, 7) and (6, 1), find the slope of a line perpendicular to l.

Solution: The slope of the line connecting the two given points is

$m = \dfrac{1-7}{6-(-4)} = \dfrac{-6}{10} = \dfrac{-3}{5}$,

so the slope of a line perpendicular to that one will be $\dfrac{5}{3}$.

Example 6

Find the equation of a line that passes through the point $(1, 1)$ and is perpendicular to $3x - 4y = 12$.

Solution: First, you need to find the slope of the given line. The simplest way to do that is to transform the equation to **slope-intercept**, or $y = mx + b$, form.

$$3x - 4y = 12 \Rightarrow -4y = -3x + 12$$

The slope of this line is $\dfrac{3}{4}$, so any line perpendicular to it will have a slope of $-\dfrac{4}{3}$. Use **point-slope** form to create the equation.

$$y - y_1 = m(x - x_1)$$
$$y - 1 = -\frac{4}{3}(x - 1)$$
$$y - 1 = -\frac{4}{3}x + \frac{4}{3}$$
$$y = -\frac{4}{3}x + \frac{7}{3}$$
$$3y = -4x + 7$$
$$4x + 3y = 7$$

Lesson 2-8 Review

1. Find the length of $\overline{AB}$ if A is the point $(2, 3)$ and B is the point $(-1, -1)$.

2. You want to draw a segment from the origin to a point whose x-coordinate is 5, and you want the segment to have a length of 13. What should be the y-coordinate of the second point?

3. Find the midpoint of the segment whose endpoints are $(4, -3)$ and $(8, 5)$.

4. If $\overline{CD}$ has endpoints $C\ (1, 6)$ and $D\ (x, 2)$ and the midpoint of $\overline{RT}$ is $(4, 4)$, find x.

5. Find the equation of a line that passes through the point $(3, 5)$ and is perpendicular to $2x + y = 4$.

Answer Key

Lesson 2-2

1. $|32 - (-15)| = 47$

2. $|r - 2.3| = 19.5 \Rightarrow r - 2.3 = \pm 19.5 \Rightarrow r = 21.8, r = -17.2$

3.

Statement	Reason
1. $AB = YZ$ $AW = CZ$	1. Given
2. $AW = AB + BW$ $CZ = CY + YZ$	2. Segment Addition
3. $AB + BW = CY + YZ$	3. Substitution
4. $BW = CY$	4. Subtraction Property of Equality

Figure 2.39

4.

Statement	Reason
1. $BW = CY$	1. Given
2. $BW = BC + CW$ $CY = CW + WZ$	2. Segment Addition
3. $BC + CW = CW + WZ$	3. Substitution
4. $CW = CW$	4. Identity
5. $BC = WY$	5. Subtraction Property of Equality

Figure 2.40

Lesson 2-3

1. $m\angle BEC = 15°$

2. $m\angle AEC = |73° - 171°| = 98°$

3.

Statement	Reason
1. m∠ZYW = m∠XWY m∠ZWY = m∠XYW	1. Given
2. m∠ZYW + m∠XYW = m∠XWY + m∠ZWY	2. Addition Property of Equality
3. m∠ZYW + m∠XYW = m∠XYZ m∠XWY + m∠ZWY = m∠ZWX	3. Angle Addition
4. m∠ZWX = m∠XYZ	4. Substitution

Figure 2.41

Lesson 2-4

1. 14   2. 26°

Lesson 2-5

1. 165°   2. 49°   3. 63°   4. 68°

Lesson 2-6

1. False

2. True

3. False

4. True

5. m∠1 = m∠2 = m∠4 = m∠6 = 155°
    m∠3 = m∠5 = 25°

Lesson 2-7

1. 45°   2. $x = 11$

Lesson 2-8

1. $AB = 5$

2. The y-coordinate could be 12 or –12.

3. The midpoint is (6, 1)

4. $x = 7$

5. $y = \dfrac{1}{2}x + \dfrac{7}{2}$ or $x - 2y = -7$

Parallel Lines

Lesson 3-1: Transversals and Angles

A line that cuts through two or more lines is called a **transversal**. When two lines are cut by a transversal, eight angles are formed. A cluster of four angles is created when the transversal crosses the first line and another cluster of four when the transversal intersects the second line. In Figure 3.1, the first cluster is labeled with the numbers 1 through 4, and the second cluster with the numbers 5 through 8.

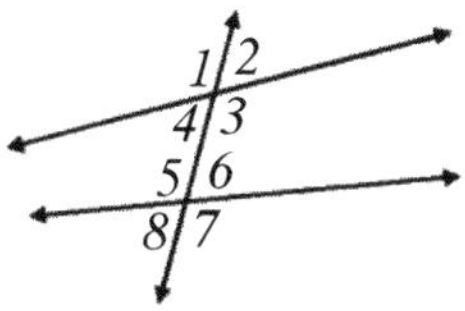

Figure 3.1

You could focus on many different pairs of angles from among the eight. For example, $\angle 1$ and $\angle 2$ form a linear pair, and $\angle 5$ and $\angle 7$ are a pair of vertical angles. Pairs of angles are named by their locations. When you talk about **corresponding angles**, you mean one angle from each cluster, in the same location within the cluster. If you start with $\angle 2$ from the top cluster, and go to the bottom cluster to look for the corresponding angle, you want the angle in the upper right of the cluster, $\angle 6$. So $\angle 2$ and $\angle 6$ are corresponding angles. Working around each cluster, you can see that $\angle 3$ and $\angle 7$ are corresponding angles, and so are $\angle 4$ and $\angle 8$, and $\angle 1$ and $\angle 5$.

The other pairs of angles all have names that involve the words *interior* or *exterior*, so let's start by talking about what those words mean. **Interior angles** are located between the two lines. By contrast, **exterior angles** are located above or below, outside the two lines. There are four interior angles, two from each cluster, and four exterior angles.

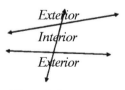

Figure 3.2

Alternate interior angles are a pair of interior angles, one from the top cluster and one from the bottom, that are on different sides of the transversal. $\angle 3$ and $\angle 5$ are alternate interior angles. There are two pairs of alternate interior angles; the other pair is $\angle 4$ and $\angle 6$. Outlining a pair of alternate interior angles will make a Z shape.

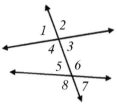

Figure 3.3

Alternate exterior angles are a pair of exterior angles, one from the top cluster and one from the bottom, that are on different sides of the transversal. As with alternate interior angles, there are two pairs of alternate exterior angles. One pair of alternate exterior angles is $\angle 1$ and $\angle 7$; the other pair is $\angle 2$ and $\angle 8$.

Interior angles on the same side of the transversal are just what their name says: They fall between the lines and they are not alternate. They sit on the same side of the transversal. $\angle 4$ and $\angle 5$ are interior angles on the same side of the transversal; so are $\angle 3$ and $\angle 6$.

Example 1

If line $\overleftrightarrow{EF}$ intersects opposite sides of quadrilateral $ABCD$ as shown in Figure 3.4, name:

a. A pair of corresponding angles.

b. A pair of alternate interior angles.

c. A pair of alternate exterior angles.

d. A pair of interior angles on the same side of the transversal.

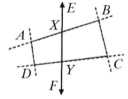

Figure 3.4

Solution:

a. To find corresponding angles, you need to find two clusters of angles where the transversal, $\overleftrightarrow{EF}$, intersects each of the two lines. One cluster is centered about X and the other about Y. Choose one angle from the topic cluster, say $\angle AXY$, and move to the bottom cluster. $\angle AXY$ is the lower left angle in its cluster, so its corresponding angle is the lower left from the bottom cluster: $\angle DYF$.

b. Look for a Z shape to find alternate interior angles. Starting from A, you can trace a Z shape by going from A to X to Y to C. One pair of alternate interior angles is $\angle AXY$ and $\angle XYC$.

c. To locate alternate exterior angles, you can start from the alternate interior angles you just found and then jump over the lines $\overleftrightarrow{AB}$ and $\overleftrightarrow{DC}$. From ∠*AXY*, jump up over $\overleftrightarrow{AB}$ to ∠*AXE*. From ∠*XYC*, jump down below $\overleftrightarrow{DC}$ to ∠*FYC*. So ∠*AXE* and ∠*FYC* are a pair of alternate exterior angles.

d. One pair of interior angles on the same side of the transversal would be ∠*AXY* and ∠*XYD*. Another pair, also using $\overleftrightarrow{EF}$ as the transversal, would be ∠*BXY* and ∠*XYC*, but notice that the sides of the quadrilateral can be extended (the dotted lines) and used as transversals too. So you might find that ∠*XAD* and ∠*ADY* are a pair of interior angles on the same side of transversal $\overleftrightarrow{AD}$.

Lesson 3-1 Review

1. Identify each of the following in Figure 3.5:
   a. A pair of alternate interior angles
   b. A pair of vertical angles
   c. A pair of corresponding angles
   d. A linear pair
   e. A pair of interior angles on the same side of the transversal
   f. A pair of alternate exterior angles

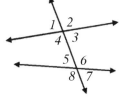

Figure 3.5

Answer True or False for questions 2–5.

___ 2. ∠4 and ∠5 are alternate interior angles.

___ 3. ∠2 and ∠8 are alternate exterior angles.

___ 4. ∠3 and ∠8 are corresponding angles.

___ 5. ∠3 and ∠6 are interior angles on the same side of the transversal.

Lesson 3-2: Parallel Lines and Angle Relationships

Parallel lines are lines that never intersect, no matter how far they are extended. Parallel lines are always the same distance apart. The symbol for "is parallel to" is ‖. If line *l* is parallel to line *m*, write *l* ‖ *m*.

Figure 3.6

The names given to the different pairs of angles depend only on their location. It doesn't matter if the lines are parallel or not. If the lines are parallel, however, then some important relationships exist among their measurements.

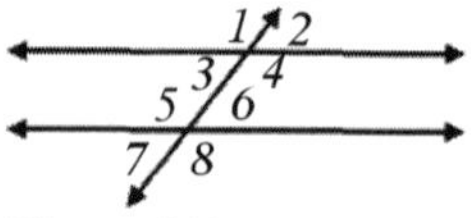

Figure 3.7

If two parallel lines are cut by a transversal:

▶ Corresponding angles are equal.

 ◆ $\angle 1 = \angle 5$, $\angle 2 = \angle 6$, $\angle 3 = \angle 7$, and $\angle 4 = \angle 8$

▶ Alternate interior angles are equal.

 ◆ $\angle 3 = \angle 6$ and $\angle 4 = \angle 5$

▶ Alternate exterior angles are equal.

 ◆ $\angle 1 = \angle 8$ and $\angle 2 = \angle 7$

▶ Interior angles on the same side of the transversal are supplementary.

 ◆ $\angle 5 + \angle 3 = 180°$ and $\angle 4 + \angle 6 = 180°$

Suppose the lines in Figure 3.8 are parallel and $m\angle 1 = 100°$. Because vertical angles are equal, $m\angle 3 = 100°$. The lines are parallel, so corresponding angles, $\angle 1$ and $\angle 5$, are equal. This means $m\angle 5 = 100°$, and by vertical angles, $m\angle 7 = 100°$. Because $\angle 1$ and $\angle 2$ are a linear pair, they are supplementary, so $m\angle 2 = 80°$. Vertical angles are equal, so $m\angle 2 = m\angle 4 = 80°$. Alternate interior angles are equal, so $m\angle 4 = m\angle 6 = 80°$, and vertical angles again tell you that $m\angle 8 = 80°$. From just knowing that the lines are parallel and that $m\angle 1 = 100°$, you now know the measurements of all eight angles.

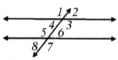

Figure 3.8

Example 1

$\overleftrightarrow{AB} \parallel \overleftrightarrow{CD}$ and transversal $\overleftrightarrow{RT}$ intersects $\overleftrightarrow{AB}$ at X and $\overrightarrow{CD}$ at Y. If $m\angle AXY = 55°$, find the measures of the other angles.

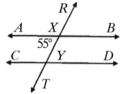

Figure 3.9

Solution: Three other angles will measure 55°. Because vertical angles are equal, and because

$\overrightarrow{AB} \| \overrightarrow{CD}$, alternate interior angles and corresponding angles are congruent, so $\angle RXB \cong \angle XYD \cong \angle CYT \cong \angle AXY$ and all of them measure 55°.

$\angle AXR$ and $\angle AXY$ are a linear pair, so they are supplementary. That means $m\angle AXR = 180 - m\angle AXY = 180° - 55° = 125°$. Because of vertical angles, and corresponding angles, $m\angle AXR = m\angle BXY = m\angle XYC = m\angle DYT = 125°$.

Example 2

$\overrightarrow{AB} \| \overrightarrow{CD}$ and transversal $\overrightarrow{RT}$ intersects $\overrightarrow{AB}$ at X and $\overrightarrow{CD}$ at Y. If $m\angle RXB = 3x - 14$, and $m\angle CYT = 126 - 4x$, find the value of x and the measure of $\angle CYX$.

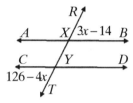

Figure 3.10

Solution: $\angle RXB$ and $\angle CYT$ are alternate exterior angles. Because $\overrightarrow{AB} \| \overrightarrow{CD}$, $m\angle RXB = m\angle CYT$. This gives you an equation to solve to find x.

$$3x - 14 = 126 - 4x$$
$$7x - 14 = 126$$
$$7x = 140$$
$$x = 20$$

Once you know the value of x, you can determine that $m\angle RXB = 3(20) - 14 = 60 - 14 = 46°$, and $m\angle CYT = 126 - 4x = 126 - 4(20) = 126 - 80 = 46°$. Because $\angle CYX$ is supplementary to $\angle CYT$, $m\angle CYT = 180 - 46 = 136°$.

Lesson 3-2 Review

1. If $\overline{BC} \| \overline{AD}$, and $m\angle BAD = x$ and $m\angle ABC = 3x + 12$, find $m\angle BAD$.

2. If $\overline{BC} \| \overline{AD}$, $m\angle ABC = 130°$, $m\angle GAF = 3x + 4$ and $m\angle FAD = 7x + 6$, find $m\angle FAD$.

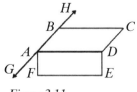

Figure 3.11

3. If $\overline{AF} \parallel \overline{DE}$ and $\overline{DE} \perp \overline{FE}$, prove $\overline{AF} \perp \overline{FE}$. *(You are proving that, if a line is perpendicular to one of two parallel lines, it is perpendicular to the other.)*

Lesson 3-3: Proving Lines Parallel

If you need to prove that two lines are parallel, you can go back to the definition and show that they are always the same distance apart, but that can be hard to do. You would have to be able to measure the perpendicular distance between the lines in two different spots, and you may not have, or be able to get, that information.

Luckily, many of the rules in Lesson 3-2 have converses that are also true. (Remember that a true conditional statement doesn't always have a true converse, so this is a fortunate circumstance.) Here is the list of all those converses, each of which tells you a way to prove that two lines are parallel:

▶ If two lines are cut by a transversal and corresponding angles are equal, then the lines are parallel.

▶ If two lines are cut by a transversal and alternate interior angles are equal, then the lines are parallel.

▶ If two lines are cut by a transversal and alternate exterior angles are equal, then the lines are parallel.

▶ If two lines are cut by a transversal and interior angles on the same side of the transversal are supplementary, then the lines are parallel.

Most often, the best way to prove that two lines are parallel is to prove that one of these angle relationships exists. That will guarantee that the lines are parallel. You only need to prove the relationship for one pair of angles; the others will fall into place.

Example 1

$\overline{EL}$ bisects both $\angle BLU$ and $\angle BEU$ and $\angle BLU \cong \angle BEU$. Prove that $\overline{BL} \parallel \overline{EU}$.

Solution: You can use the halves of equals subroutine to prove that $\angle BLE \cong \angle LEU$. Because those are alternate interior angles, if they are congruent, $\overline{BL} \parallel \overline{EU}$.

Statement	Reason
1. $\overline{EL}$ bisects $\angle BLU$ $\overline{EL}$ bisects $\angle BEU$	1. Given
2. $m\angle BLE = m\angle ELU =$ $\frac{1}{2} \, m\angle BLU$ $m\angle UEL = m\angle LEB =$ $\frac{1}{2} \, m\angle BEU$	2. An angle bisector divides the angle into two equal angles, each of which is half as large as the original.
3. $m\angle BLU = m\angle BEU$	3. Given
4. $\frac{1}{2} \, m\angle BLU =$ $\frac{1}{2} \, m\angle BEU$	4. Multiplication Property of Equality
5. $m\angle BLE = m\angle UEL$	5. Substitution
6. $\overline{BL} \parallel \overline{EU}$	6. If two lines are cut by a transversal and the alternate interior angles are congruent, then the lines are parallel.

Figure 3.12

Example 2

If $m\angle MSI = 3x + 27$, $m\angle SOR = 5x - 17$, and $m\angle RON = 4x + 8$, is it possible that $\overleftrightarrow{IS} \parallel \overleftrightarrow{RO}$? Explain your reasoning.

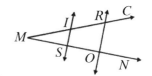

Figure 3.13

Solution: You might be tempted to set $\angle MSI$ equal to $\angle SOR$, but corresponding angles like these are only congruent if the lines are parallel, and you don't know yet if they are parallel. Whether $\overleftrightarrow{IS} \parallel \overleftrightarrow{RO}$ or not, $\angle SOR$ and $\angle RON$ form a linear pair, and so they are supplementary. Use this as basis of an equation to find the value of x.

$$m\angle SOR + m\angle RON = 180$$
$$5x - 17 + 4x + 8 = 180$$
$$9x - 9 = 180$$
$$9x = 189$$
$$x = 21$$

Once you know the value of x, you can determine the measure of each angle. $m\angle MSI = 3(21) + 27 = 63 + 27 = 90°$, $m\angle SOR = 5(21) - 17 = 105 - 17 = 88°$, and $m\angle RON = 4(21) + 8 = 84 + 8 = 92°$.

Once you know the measure of each angle, you can see that $\angle MSI$ is not equal to $\angle SOR$. Because these corresponding angles are not equal, the lines could not possibly be parallel.

Lesson 3-3 Review

1. If $\angle FRA = 4x + 5$, $\angle ARD = 10x - 18$, $\angle ERD = 4x - 5$, and $\angle DAR$ and $\angle ADR$ are complementary, is it possible that $\overline{AD} \parallel \overline{FE}$? Explain your reasoning.

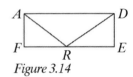

Figure 3.14

2. If $\overline{AF} \perp \overline{FE}$ and $\overline{DE} \perp \overline{FE}$, prove $\overline{AF} \parallel \overline{DE}$.

   (You are proving that, if two lines are perpendicular to the same line, they are parallel to each other.)

Lesson 3-4: The Angles of a Triangle

In an earlier math class, you may have conducted some experiments to convince yourself that the three angles of a triangle add to 180°, perhaps by measuring the angles of several different triangles with a protractor and checking to see that the three angles always add to 180°. You might have clipped the corners off a triangle, and rearranged them to form a straight angle. Such experiments help you to understand and to remember the relationship, but they don't constitute proof.

You know almost enough now to construct a proof of that theorem. It will be similar to snipping off the corners and rearranging them to form a straight angle, but instead of snipping and rearranging, you will use what you know about parallel lines and angle relationships. You just need one more postulate.

Given a line and a point not on the line, there is exactly one line through the given point that is parallel to the given line.

Because this is a postulate, it is accepted without proof. The postulate says that you can draw a line through the point that will be parallel to the line, and you can draw only one line that way.

With that postulate, you are ready to prove that the angles of a triangle add to 180°. A diagram will help illustrate the situation, but remember that you are not proving this for one particular triangle but for all triangles.

Given $\triangle ABC$, draw a line through B, parallel to $\overline{AC}$, and call it $\overleftrightarrow{DE}$. You know, because of the postulate, that there is such a line—and there is only one such line.

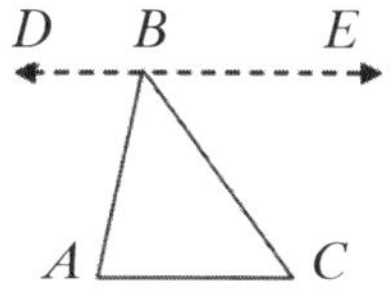

Figure 3.15

Taken together, $\angle DBA$, $\angle ABC$, and $\angle CBE$ form a straight angle, so m$\angle DBA$ + m$\angle ABC$ + m$\angle CBE$ = 180°. You know that, when parallel lines are cut by a transversal, alternate interior angles are congruent, so, using $\overline{AB}$ as the transversal, you can say $\angle DBA \cong \angle BAC$, and using $\overline{BC}$ as the transversal, $\angle CBE \cong \angle ACB$. You can substitute, so m$\angle DBA$ + m$\angle ABC$ + m$\angle CBE$ = 180° becomes m$\angle BAC$ + m$\angle ABC$ + m$\angle ACB$ = 180°, which was what you wanted.

The assurance that the three angles of any triangle always add to 180° is helpful in problem-solving and in proofs.

Example 1

Given $\angle RAD = 7x + 6$, $\angle ADR = 5x$, and $\angle ARD = 13x - 1$. Are $\angle FRA$ and $\angle ERD$ complementary?

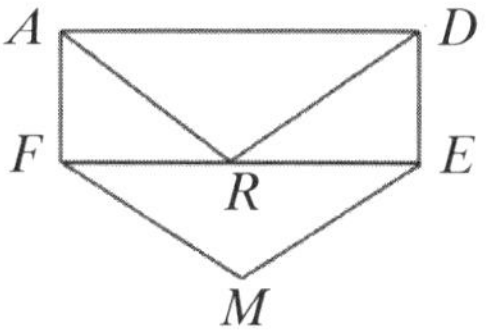

Solution: $\angle RAD$, $\angle ADR$, and $\angle ARD$ are the angles of a triangle, so they must add to 180°.

Figure 3.16

$$m\angle RAD + m\angle ADR + m\angle ARD = 180°$$
$$7x + 6 + 5x + 13x - 1 = 180$$
$$25x + 5 = 180$$
$$25x = 175$$
$$x = 7$$

m$\angle RAD$ = 7(7) + 6 = 55°, m$\angle ADR$ = 5(7) = 35°, and m$\angle ARD$ = 13(7) − 1 = 90°. Because $\angle ARD$ is a right angle, and because $\angle ARD$, $\angle FRA$, and $\angle ERD$ together form a straight angle, you know m$\angle ARD$ + m$\angle FRA$ + m$\angle ERD$ = 180. Substituting, 90 + m$\angle FRA$ + m$\angle ERD$ = 180 gives you m$\angle FRA$ + m$\angle ERD$ = 90. Therefore, $\angle FRA$ and $\angle ERD$ are complementary.

Example 2

Given: $\angle RFA \cong \angle EMF$ and $\angle FAR \cong \angle FEM$.

Prove: $\overline{AR} \parallel \overline{FM}$

Solution:

Statement	Reason
1. $m\angle MFE + m\angle FEM + m\angle EMF = 180°$ $m\angle FAR + m\angle ARF + m\angle RFA = 180°$	1. The three angles of a triangle add to 180°.
2. $m\angle MFE + m\angle FEM + m\angle EMF = m\angle FAR + m\angle ARF + m\angle RFA$	2. Transitive Property of Equality
3. $\angle RFA \cong \angle EMF$ and $\angle FAR \cong \angle FEM$	3. Given
4. $m\angle RFA = m\angle EMF$ and $m\angle FAR = m\angle FEM$	4. Congruent angles have equal measures.
5. $m\angle MFE = m\angle ARF$	5. Subtraction Property of Equality
6. $\overline{AR} \parallel \overline{FM}$	6. If two lines are cut by a transversal and the alternate interior angles are congruent, then the lines are parallel.

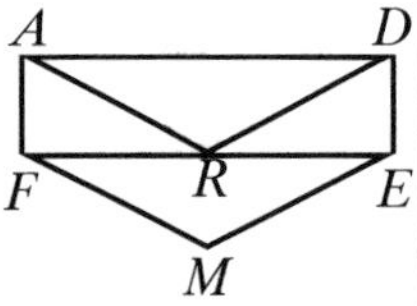

Figure 3.17

Lesson 3-4 Review

Use Figure 3.18.

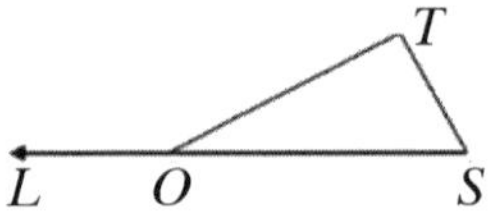

1. If $m\angle TOS = 34°$ and $m\angle OTS = 80°$, find $m\angle TSO$.

Figure 3.18

2. If $m\angle TOS = 2x$, $m\angle OTS = 7x$ and $m\angle TSO = 3x$, find $m\angle LOT$.

3. If $m\angle LOT = 130°$ and $m\angle OTS = 75°$, find $m\angle TSO$.

4. Prove that $m\angle LOT = m\angle OTS + m\angle TSO$.

Lesson 3-5: Parallel Lines in the Coordinate Plane

In most cases, if you want to prove lines parallel, you prove a pair of angles congruent. In coordinate geometry, you can use slope to help understand the angles that lines make. In Lesson 2-8, you saw that perpendicular lines, which form right angles, have slopes that are negative reciprocals. Parallel lines must make the same angles with their transversal. Because the angles must be equal, the slopes, which measure angles, must be equal. Therefore, parallel lines have the same slope.

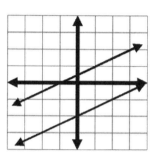

Figure 3.19

Example 1

Line k passes through the points (4, 3) and (−1, 8). Line l passes through the points (−4, −3) and (8, 3). Are the lines parallel, perpendicular, or neither?

Solution: Line k has a slope of $m_k = \dfrac{3-8}{4-(-1)} = \dfrac{-5}{5} = -1$.

Line l has a slope of $m_l = \dfrac{-3-3}{-4-8} = \dfrac{-6}{-12} = \dfrac{1}{2}$.

Because the slopes are not the same, the lines are not parallel, and because the product of the slopes is $-\dfrac{1}{2}$, not −1, the lines are not perpendicular. So the lines are neither parallel nor perpendicular.

Example 2

Are the lines $3x - 6y = 9$ and $8y = 12 + 4x$ parallel, perpendicular, or neither?

Solution: Determine the slope of each line by transforming each equation to slope-intercept form.

$$3x - 6y = 9 \Rightarrow -6y = -3x + 9 \Rightarrow y = \frac{-3}{-6}x - \frac{9}{6} \Rightarrow y = \frac{1}{2}x - \frac{3}{2}$$

and $8y = 12 + 4x \Rightarrow y = \dfrac{12}{8} + \dfrac{4}{8}x \Rightarrow y = \dfrac{1}{2}x + \dfrac{3}{2}$

The slope of each line is $\dfrac{1}{2}$. Because the lines have the same slope, they are parallel.

Example 3

Find a line through the point $(-7, 5)$ parallel to the line $y = \dfrac{3}{7}x - 1$.

Solution: The slope of the given line is $\dfrac{3}{7}$. To find the equation of the line with slope $\dfrac{3}{7}$ that passes through the point $(-7, 5)$, use point-slope form.

$$y - 5 = \frac{3}{7}\left(x - (-7)\right)$$

$$y - 5 = \frac{3}{7}x + 3$$

$$y = \frac{3}{7}x + 8$$

The line $y = \dfrac{3}{7}x + 8$ passes through $(-7, 5)$ and is parallel to $y = \dfrac{3}{7}x - 1$.

Lesson 3-5 Review

1. Are the lines $5x - 3y = -15$ and $2x - 6y = 18$ parallel, perpendicular, or neither?

2. Find a line through the point $(8, -3)$ parallel to the line $2x - 5y = 7$.

Answer Key
Lesson 3-1

1. a. $\angle 4$ and $\angle 6$, or $\angle 3$ and $\angle 5$
   b. $\angle 1$ and $\angle 3$, or $\angle 2$ and $\angle 4$, or $\angle 5$ and $\angle 7$, or $\angle 6$ and $\angle 8$
   c. $\angle 1$ and $\angle 5$, or $\angle 2$ and $\angle 6$, or $\angle 3$ and $\angle 7$, or $\angle 4$ and $\angle 8$

d. $\angle 1$ and $\angle 2$, or $\angle 2$ and $\angle 3$, or $\angle 3$ and $\angle 4$, or $\angle 4$ and $\angle 1$, or $\angle 5$ and $\angle 6$, or $\angle 6$ and $\angle 7$, or $\angle 7$ and $\angle 8$, or $\angle 8$ and $\angle 5$

e. $\angle 3$ and $\angle 6$, or $\angle 4$ and $\angle 5$

f. $\angle 1$ and $\angle 7$, or $\angle 2$ and $\angle 8$

2.   False   3.   True   4.   False   5.   True

Lesson 3-2

1. $m\angle BAD + m\angle ABC = 180$

$$x + 3x + 12 = 180$$
$$4x + 12 = 180$$
$$4x = 168$$
$$x = 42$$

2. $m\angle ABC = m\angle GAD$

$m\angle ABC = m\angle GAF + m\angle FAD$

$$130 = 3x + 4 + 7x + 6$$
$$130 = 10x + 10$$
$$120 = 10x$$
$$12 = x$$

$m\angle FAD = 7(10) + 6 = 76°$

3.

Statement	Reason
1. $\overline{AF} \parallel \overline{DE}$	1. Given
2. $\angle AFE$ and $\angle DEF$ are supplementary.	2. If parallel lines are cut by a transversal, interior angles on the same side of the transversal are supplementary.
3. $m\angle AFE + m\angle DEF = 180°$	3. If two angles are supplementary, their measures total 180°.
4. $\overline{DE} \perp \overline{FE}$	4. Given
5. $\angle DEF$ is a right angle.	5. If two lines are perpendicular, they intersect to form right angles.
6. $m\angle DEF = 90°$	6. If an angle is a right angle, it measures 90°.
7. $m\angle AFE + 90° = 180°$	7. Substitution
8. $m\angle AFE = 90°$	8. Subtraction Property of Equality
9. $\angle AFE$ is a right angle.	9. If an angle measures 90°, it is a right angle.
10. $\overline{AF} \perp \overline{FE}$	10. If two lines intersect to form a right angle, they are perpendicular.

Figure 3.20

Lesson 3-3

1. $m\angle FRA + m\angle ARD + m\angle ERD = 180$

$$4x + 5 + 10x - 18 + 4x - 5 = 180$$
$$18x - 18 = 180$$
$$18x = 198$$
$$x = 11$$

$\angle FRA = 4(11) + 5 = 49°, \angle ARD = 10(11) - 18 = 92°, \angle ERD = 4(11) - 5 = 39°.$

$\overline{AD}$ is not parallel to $\overline{FE}$.

2.

Statement	Reason
1. $\overline{AF} \perp \overline{FE}$ $\overline{DE} \perp \overline{FE}$	1. Given
2. $\angle AFE$ is a right angle. $\angle DEF$ is a right angle.	2. If two lines are perpendicular, they intersect to form right angles.
3. $m\angle AFE = 90°$ $m\angle DEF = 90°$	3. If an angle is a right angle, it measures 90°.
4. $m\angle AFE + m\angle DEF = 180°$	4. Addition Property of Equality
5. $\angle AFE$ and $\angle DEF$ are supplementary.	5. If the measures of two angles add to 180°, the angles are supplementary.
6. $\overline{AF} \parallel \overline{DE}$	6. If two lines are cut by a transversal and interior angles on the same side of the transversal are supplementary, then the lines are parallel.

Figure 3.21

Lesson 3-4

1. $m\angle TSO = 66°$

$$m\angle TOS + m\angle OTS + m\angle TSO = 180$$
$$34 + 80 + x = 180$$
$$114 + x = 180$$
$$x = 66°$$

2. $m\angle LOT = 150°$

$$m\angle TOS + m\angle OTS + m\angle TSO = 180$$
$$2x + 7x + 3x = 180$$
$$12x = 180$$
$$x = 15$$

$m\angle TOS = 2(15) = 30°. m\angle LOT = 180 - 30 = 150°.$

3. $m\angle TSO = 55°$. If $m\angle LOT = 130°$, then $m\angle TOS = 50°$.

$$m\angle TOS + m\angle OTS + m\angle TSO = 180$$
$$50 + 75 + x = 180$$
$$125 + x = 180$$
$$x = 55°$$

4.

Statement	Reason
1. $\angle LOT$ and $\angle TOS$ are a linear pair.	1. Definition of a linear pair
2. $\angle LOT$ and $\angle TOS$ are supplementary.	2. Linear pairs are supplementary.
3. $m\angle LOT + m\angle TOS = 180°$	3. Supplementary angles add to 180°.
4. $m\angle TOS + m\angle OTS + m\angle TSO = 180°$	4. The three angles of a triangle add to 180°.
5. $m\angle LOT + m\angle TOS = m\angle TOS + m\angle OTS + m\angle TSO$	5. Transitive Property of Equality
6. $m\angle LOT = m\angle OTS + m\angle TSO$	6. Subtraction Property of Equality

Figure 3.22

Lesson 3-5

1. $5x - 3y = -15$ has a slope of $\frac{5}{3}$ and $2x - 6y = 18$ has a slope of $\frac{1}{3}$.

   The lines are neither parallel nor perpendicular.

2. $2x - 5y = 7 \Rightarrow -5y = -2x + 7 \Rightarrow y = \frac{2}{5}x - \frac{7}{5}$,

   so the line must have a slope of $\frac{2}{5}$.

$$y - (-3) = \frac{2}{5}(x - 8)$$
$$y + 3 = \frac{2}{5}x - \frac{16}{5}$$
$$y = \frac{2}{5}x - \frac{31}{5}$$
$$5y = 2x - 31$$
$$2x - 5y = 31$$

Congruent Triangles

Lesson 4-1: Polygons and Congruence

The word *congruent* is often used to mean "have the same measure." Two segments are congruent if they are the same length, and two angles are congruent if they have equal degree measures. To say that two polygons are congruent you must be able to establish a correspondence between their vertices so that corresponding parts are congruent. You must find a way to match up the vertices, or corners, of the figures so that the angles at matching corners have the same measurements and the sides connecting matching vertices have the same length. Visually, congruent figures are identical, except for the letters labeling their vertices and possibly their position.

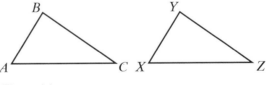

Figure 4.1

When the polygons are oriented the same way, it can be easy to see what matches. In the triangles in Figure 4.1, the angle at A is congruent to the angle at X, $\angle B \cong \angle Y$ and $\angle C \cong \angle Z$. Side $\overline{AB}$ is congruent to side $\overline{XY}$, $\overline{BC} \cong \overline{YZ}$, and $\overline{CA} \cong \overline{ZX}$. That correspondence can be summarized by writing $\triangle ABC \cong \triangle XYZ$. The order in which you name the vertices tells what vertices correspond. In $\triangle ABC \cong \triangle XYZ$, A is named first, so it corresponds to X, which is named first in its triangle. B corresponds to Y because both are named second, and so C corresponds to Z. The congruence statement must tell which parts are congruent.

When the orientation of the polygons is different, it may be harder to see the correspondence. Look for the smallest, or the largest, angle or side in each triangle first. Aligning those will help you see what else corresponds. Tracing one polygon and laying it over the other may help you find the matching parts. In the triangles in Figure 4.2, the smallest angle in the first triangle is ∠F and the smallest angle in the second triangle is ∠G, so those should correspond. It is a little harder to decide whether ∠R matches ∠E or ∠T, but ∠R is larger than ∠O, and ∠T is larger than ∠E, so the correspondence should be ΔFOR ≅ ΔGET.

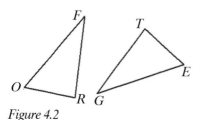

Figure 4.2

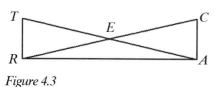

Figure 4.3

If the polygons overlap one another, it may be easier to see the correspondence when they are separated, so consider tracing the two polygons in different locations on a sheet of paper. Figure 4.3 contains several triangles, some of which overlap each other. It may be easier to determine if the overlapping triangles are congruent when they are drawn separately, as in Figure 4.4.

Point E, which might have been a distraction, becomes unnecessary when the triangles are separated, and it becomes easier to see that ΔTRA ≅ ΔCAR.

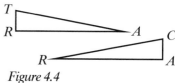

Figure 4.4

Example 1

Are the polygons congruent? If so, write the congruence statement. If not, explain why.

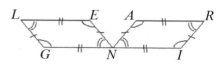

Figure 4.5

Solution: There are two possible correspondences. One can be illustrated by imagining a vertical fold in the page at N. The quadrilaterals would align with R matching L, A matching E, I matching G, and the N vertex from each quadrilateral matching. Checking angles, you see that ∠R ≅ ∠L, ∠A ≅ ∠E, ∠I ≅ ∠G, and ∠ANI ≅ ∠ENG. Checking sides, you find that $\overline{RA} ≅ \overline{LE}$,

$\overline{RI} \cong \overline{LG}$, $\overline{IN} \cong \overline{GN}$, and $\overline{AN} \cong \overline{EN}$. Because corresponding parts are congruent, $LENG \cong RANI$. (The other possibility can be visualized by flipping $NAIR$ top to bottom and results in the correspondence $LENG \cong NIRA$.)

Example 2

Are the triangles congruent? If so, write the congruence statement. If not, explain why.

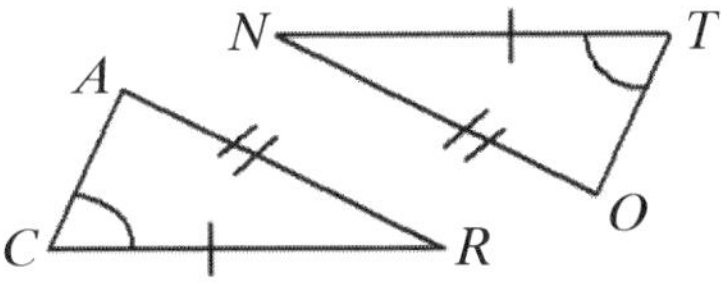

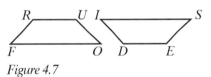

Figure 4.6

Solution: Although the visual impression may be that the triangles are the same, you can't trust your eye for measurements. You only know for certain that $\angle C \cong \angle T$, $\overline{CR} \cong \overline{TN}$, and $\overline{AR} \cong \overline{ON}$, and that is not sufficient information to be certain that the triangles are congruent.

Once you know that two polygons are congruent, you are assured that corresponding parts of the polygons are congruent. If quadrilateral $FOUR$ is congruent to quadrilateral $SIDE$, the congruence statement $FOUR \cong SIDE$ tells you the correspondence be-tween vertices. F corresponds to S, O to I, and so on. From that you can conclude that $\angle F \cong \angle S$, $\angle O \cong \angle I$, $\angle U \cong \angle D$, $\angle R \cong \angle E$, and that $\overline{FO} \cong \overline{SI}$, $\overline{OU} \cong \overline{ID}$, $\overline{UR} \cong \overline{DE}$, and $\overline{RF} \cong \overline{ES}$. For congruent triangles, this rule is usually stated as:

Figure 4.7

Corresponding parts of congruent triangles are congruent (and often abbreviated as CPCTC).

Example 3

$\triangle MUD \cong \triangle PIE$. List all the congruent angles and segments.

Solution: Follow the letters in the order they are named. $\angle M \cong \angle P$, $\angle U \cong \angle I$, $\angle D \cong \angle E$, $\overline{MU} \cong \overline{PI}$, $\overline{UD} \cong \overline{IE}$, and $\overline{DM} \cong \overline{EP}$.

Example 4

SLOPE and *GRAPH* are pentagons, and *SLOPE* ≅ *GRAPH*. List all the congruent angles and segments.

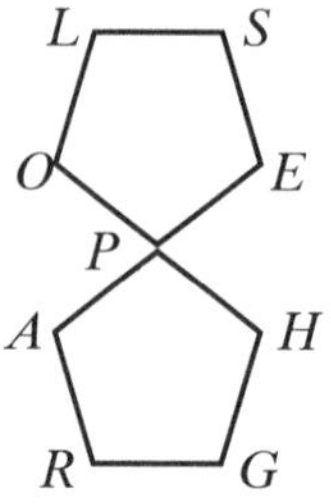

Solution: It is possible to answer the question without a diagram, just by following the correspondence given in the statement *SLOPE* ≅ *GRAPH*, but if it helps you, sketch a picture. ∠*S* ≅ ∠*G*, ∠*L* ≅ ∠*R*, ∠*O* ≅ ∠*A*, ∠*OPE* ≅ ∠*APH*, ∠*E* ≅ ∠*H*, $\overline{SL} \cong \overline{GR}$, $\overline{LO} \cong \overline{RA}$, $\overline{OP} \cong \overline{AP}$, $\overline{PE} \cong \overline{PH}$, and $\overline{ES} \cong \overline{HG}$.

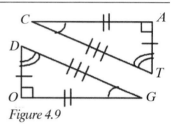

Figure 4.8

Lesson 4-1 Review

1. Are the triangles congruent? If so, write the congruence statement. If not, explain why.

2. Δ*ARM* ≅ Δ*LEG*. List all the congruent angles and segments.

Figure 4.9

Lesson 4-2: Congruence Shortcuts

The rest of this chapter will look only at congruent triangles and shortcuts for proving triangles congruent. It is not necessary to prove that all three angles and all three sides of one triangle are congruent to their corresponding parts in the other triangle. There are four minimum standards for proving any two triangles congruent. Each one gives you a list of three things you must prove, and, if you can do that, you can be certain the triangles are congruent (and the other corresponding parts of the triangles will be congruent as well). The rules are carefully worded and very specific. To use them, you must prove exactly the pieces the postulate or theorem lists.

    SSS: If three sides of one triangle are congruent to the corresponding sides of another triangle, then the triangles are congruent.

    SAS: If two sides and the included angle of one triangle are congruent to the corresponding parts of another triangle, then the triangles are congruent.

ASA: If two angles and the included side of one triangle are congruent to the corresponding parts of another triangle, then the triangles are congruent.

AAS: If two angles and a non-included side of one triangle are congruent to the corresponding parts of another triangle, then the triangles are congruent.

The postulates (SSS, SAS, and ASA) are accepted without proof. Each of them fixes the size and shape of the triangle so that there is only one possible way to finish it. SAS, for example, fixes the lengths of two sides and the angle they make with one another. The only way to complete a triangle is to connect the ends of those two segments. Only one triangle has these sides joined at this angle. Any triangle that meets these requirements is identical to every other triangle that meets them.

Figure 4.10

The last item in the list (AAS) is a theorem. Its proof uses the fact that the three angles of any triangle add to 180° to show that, whenever you have AAS, you also have ASA. After you prove the AAS theorem, you can use it just the way you would use the postulates.

Be careful not to invent your own short-cuts. Other combinations, such as SSA and AAA, will not guarantee congruent triangles. In Figure 4.11, $\overline{AB} \cong \overline{XY}$, $\overline{AC} \cong \overline{XZ}$, and $\angle B \cong \angle Y$, but you can clearly see that $\triangle ABC$ is not congruent to $\triangle XYZ$.

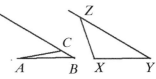

Figure 4.11

Example 1

Which postulate or theorem can be used to prove the triangles congruent?

Solution: According to the markings in Figure 4.12, there are two pairs of congruent sides and the angles included between those sides are also congruent. The triangles are congruent by SAS.

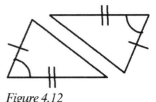

Figure 4.12

Example 2

Which postulate or theorem can be used to prove
the triangles congruent?

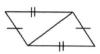

Figure 4.13

Solution: The markings on the diagram give two pairs
of congruent sides, which does not at first seem to
be enough. But the two triangles share a side, and that side is
congruent to itself, so the triangles are congruent by SSS.

Proving that two triangles are congruent is a common exercise on its
own and a common stage in a longer proof. Always examine the two
triangles before you begin and make sure you know which of the shortcuts you will use to show that they are congruent. Mark the diagram to
show which sides and angles are congruent. When you start writing the
proof, keep your shortcut in mind. There are three pieces to each shortcut. Do them in the order they are listed, and, if your teacher doesn't
object, mark each piece with a letter to signal to you what part of the
shortcut it is. Your reasons for claiming each congruence will vary, so
the reasons have been left out here.

Subroutines

SAS

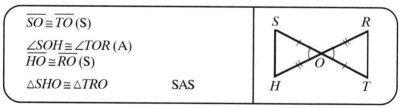

Figure 4.14

SSS

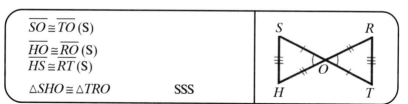

Figure 4.15

ASA

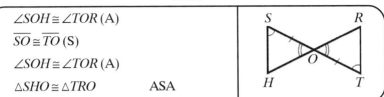

$\angle SOH \cong \angle TOR$ (A)	
$\overline{SO} \cong \overline{TO}$ (S)	
$\angle SOH \cong \angle TOR$ (A)	
$\triangle SHO \cong \triangle TRO$	ASA

Figure 4.16

Example 3

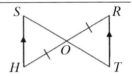

Given: $\overline{HS} \parallel \overline{TR}$ and $\overline{HO} \cong \overline{OR}$

Prove: $\triangle SHO \cong \triangle TRO$

Figure 4.17

Solution: The parallel lines can help you find congruent angles, and you also have vertical angles. Prove the triangles congruent by ASA.

Statement	Reason
1. $\overline{HS} \parallel \overline{TR}$	1. Given
2. $\angle SOH \cong \angle TOR$ (A)	2. If parallel lines are cut by a transversal, alternate interior angles are congruent.
3. $\overline{HO} \cong \overline{RO}$ (S)	3. Given
4. $\angle SOH \cong \angle TOR$ (A)	4. Vertical angles are congruent.
5. $\triangle SHO \cong \triangle TRO$	5. ASA

Figure 4.18

Example 4

Given: $\overline{IL} \cong \overline{DA}$, $\overline{IS} \cong \overline{NA}$, and $\overline{NL} \cong \overline{DS}$,

Prove: $\triangle ISL \cong \triangle AND$.

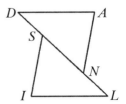

Solution: If you separate the triangles, you can see that you are given SS, but only part of each of the third sides. Use the Add Segments subroutine to prove $\overline{SL} \cong \overline{DN}$, and then prove the triangles congruent by SSS.

Figure 4.19

Statement	Reason
1. $\overline{IL} \cong \overline{DA}$ (S)	1. Given
2. $\overline{IS} \cong \overline{NA}$ (S)	2. Given
3. $\overline{NL} \cong \overline{DS}$	3. Given
4. $NL = DS$	4. Congruent segments have equal length
5. $SN = SN$	5. Reflexive
6. $SN + NL = DS + SN$	6. Addition Property of Equality
7. $SL = SN + NL$	7. Segment Addition
8. $DN = DS + SN$	8. Segment Addition
9. $SL = DN$	9. Substitution
10. $\overline{SL} \cong \overline{DN}$ (S)	10. Congruent segments have equal length
11. $\triangle ISL \cong \triangle AND$	11. SSS

Figure 4.20

Lesson 4-2 Review

1. If the triangles can be proved congruent, name the postulate or theorem that justifies the congruence. Otherwise, write "not congruent."

   a. 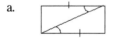   b.   c.

Figure 4.21

2. **Given:** $\overline{WH} \perp \overline{WA}$, $\overline{EL} \perp \overline{LA}$, $\overline{WA} \cong \overline{LA}$, and $\angle WAE \cong \angle LAH$

   Prove: $\triangle WHA \cong \triangle LEA$.

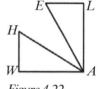

Figure 4.22

3. **Given:** $\overline{DO} \cong \overline{LP}$, $\angle ILO \cong \angle IOL$, and $\angle N \cong \angle H$,

   Prove: $\triangle DLN \cong \triangle POH$.

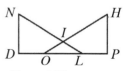

Figure 4.23

Lesson 4-3: Isosceles and Equilateral Triangles

When triangles are classified by the number of congruent sides they have, a triangle that has all three sides of different lengths is called a **scalene triangle**. A triangle with two sides of equal length is an **isosceles triangle**, and a triangle with all sides of equal length is an **equilateral triangle**. In an isosceles triangle, the two equal sides are called the legs, and the third side is the base. The base angles are the angles formed where each leg meets the base. The remaining angle, formed by the two equal sides, is the vertex angle. In Figure 4.24, the legs are $\overline{KS}$ and $\overline{KY}$, the base is $\overline{SY}$, and the base angles are $\angle S$ and $\angle Y$. The vertex angle is $\angle K$.

Figure 4.24

Commonly stated as "base angles of an isosceles triangle are congruent," the theorem says that, if two sides of a triangle are congruent, then the angles opposite those sides are congruent. The statements are equivalent because the base angles are opposite the congruent legs.

Figure 4.25

There are two ways to prove the theorem. The given is simply a triangle with two equal sides, and you are asked to prove two angles congruent. You may not feel that you can do much with only one triangle, and you would be right, so each method has a trick that produces two triangles that you can prove congruent.

In the first method, you make a copy of the triangle, and prove one copy congruent to the other, but with a twist in the correspondence. You prove $\triangle SKY \cong \triangle YKS$ by SSS, with

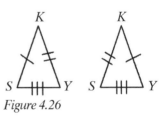

Figure 4.26

$\overline{KS} \cong \overline{KY}$ (S), $\overline{KY} \cong \overline{KS}$ (S), and $\overline{SY} \cong \overline{SY}$ (S). By CPCTC, $\angle S \cong \angle Y$.

In the second method, you draw the bisector of $\angle SKY$. This creates two triangles that can be proved congruent.

Given: $\triangle SKY$ with $\overline{KS} \cong \overline{KY}$

Prove: $\angle S \cong \angle Y$

Plan: Draw $\overrightarrow{KE}$, the bisector of $\angle SKY$, which intersects $\overline{SY}$ at E, to create $\triangle SEK$ and $\triangle YEK$. Prove the triangles congruent by SAS. By CPCTC, $\angle S \cong \angle Y$.

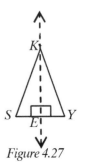

Figure 4.27

Statement	Reason
1. △SKY	1. Given
2. Draw $\overline{KE}$, the bisector of ∠SKY	2. An angle has exactly one bisector.
3. $\overline{KE} \cong \overline{KE}$ (S)	3. Reflexive
4. ∠SKE ≅ ∠YKE (A)	4. A bisector divides the angle into two congruent angles.
5. $\overline{KS} \cong \overline{KY}$ (S)	5. Given
6. △SEK ≅ △YEK	6. SAS
7. ∠S ≅ ∠Y	7. CPCTC

Figure 4.28

This theorem can be extended to equilateral triangles to show that every equilateral triangle is equiangular. In equilateral triangle △ABC,

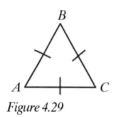

Figure 4.29

because $\overline{AB} \cong \overline{BC}$, ∠A ≅ ∠C. Because $\overline{AB} \cong \overline{AC}$, ∠B ≅ ∠C. Applying the symmetric and transitive properties, we see that if ∠A ≅ ∠C and ∠C ≅ ∠B, then ∠A ≅ ∠B, and all three angles are congruent. (It is not difficult to combine this theorem with the one that tells you the three angles of a triangle add to 180° to prove that each of these congruent angles measures 60°.)

Each of these theorems tells you that if you have congruent sides, there will also be congruent angles. The converse of each theorem is also true: If two angles of a triangle are congruent, the sides opposite those angles are congruent, and every equiangular triangle is equilateral.

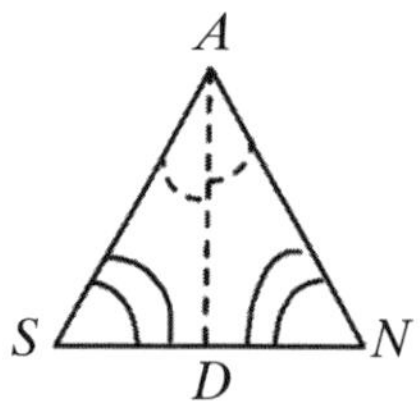

Figure 4.30

Given a triangle as shown in Figure 4.30 with two congruent angles (∠S ≅ ∠N), draw the bisector of the third angle (∠A). This creates two triangles that you can prove congruent by AAS. One set of congruent angles is given (∠S ≅ ∠N), and the bisector creates two congruent angles (∠SAD ≅ ∠NAD). The bisector segment ($\overline{AD}$) congruent to itself, gives you the side in AAS. CPCTC assures that the sides opposite the congruent angles ($\overline{AS}$ and $\overline{AN}$) are congruent sides.

Example 1

In $\triangle SUN$, $m\angle S = 5x - 13$, $m\angle U = 4x + 5$, and $m\angle N = 2x - 10$. Is $\triangle SUN$ equilateral, isosceles, or scalene?

Solution: The three angles of the triangle add to 180° so:

$$m\angle S + m\angle U + m\angle N = 180°$$
$$5x - 13 + 4x + 5 + 2x - 10 = 180$$
$$11x - 18 = 180$$
$$11x = 198$$
$$x = 18$$

If $x = 18$, $m\angle S = 5(18) - 13 = 90 - 13 = 77°$,
$m\angle U = 4(18) + 5 = 72 + 5 = 77°$, and
$m\angle N = 2(18) - 10 = 36 - 10 = 26°$.

Because two angles in the triangle are congruent ($\angle S \cong \angle U$), the sides opposite these angles are congruent.

That means $\overline{UN} \cong \overline{SN}$ and the triangle is isosceles.

Example 2

If $\triangle TRS$ is equilateral and $\overline{TR} \parallel \overline{SE}$, prove that $\overline{SR}$ bisects $\angle TSE$.

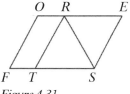

Figure 4.31

Solution: To show that $\angle TSE$ is bisected by $\overline{SR}$, you must show that $\angle TSR \cong \angle RSE$. The parallel lines allow you to show that alternate interior angles, $\angle RSE$ and $\angle TRS$, are congruent, and the equilateral triangle is equiangular, so you can show $\angle TSR \cong \angle RSE$ by transitivity.

Statement	Reason
1. $\overline{TR} \parallel \overline{SE}$	1. Given
2. $\angle RSE \cong \angle TRS$	2. If parallel lines are cut by a transversal, alternate interior angles are congruent.
3. $\triangle TRS$ is equilateral.	3. Given
4. $\triangle TRS$ is equiangular.	4. Every equilateral triangle is equiangular.
5. $\angle TRS \cong \angle TSR \cong \angle RTS$	5. An equilateral triangle has three congruent angles.
6. $\angle TSR \cong \angle RSE$	6. Transitive Property
7. $\overline{SR}$ bisects $\angle TSE$	7. A segment that divides an angle into two congruent angles is an angle bisector.

Figure 4.32

Lesson 4-3 Review

Use Figure 4.33 for questions 1 and 2.

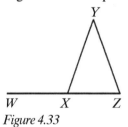

Figure 4.33

1. If $\triangle XYZ$ is isosceles with $\overline{XY} \cong \overline{YZ}$, and m$\angle WXY = 112°$, find the measure of $\angle XYZ$.

2. If $\triangle XYZ$ is isosceles with $\overline{XY} \cong \overline{YZ}$, m$\angle YXZ = 5x + 3$ and m$\angle YZX = 111 - 4x$, find the measure of $\angle XYZ$ and the measure of $\angle WXY$.

3. $\triangle LAK$ is a right triangle with right angle at A. $\overline{LE} \cong \overline{AE} \cong \overline{EK}$. Prove $\angle L$ and $\angle K$ are complementary.

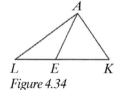

Figure 4.34

Lesson 4-4: HL Congruence

One other shortcut for proving triangles congruent, called hypotenuse-leg, or HL, may seem even shorter than the other postulates and theorems. It sounds as though there are only two things to prove instead of three, but this shortcut applies only to right triangles. In fact, you must prove three things: that both triangles are right triangles, that the hypotenuse of the first is congruent to that of the second, and that a pair of corresponding legs is congruent.

It is essential to demonstrate that the triangles are right triangles, as your evidence that you have the right to use HL to prove the triangles congruent.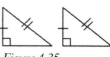

Figure 4.35

To prove that the triangles are right triangles, you must show that they contain right angles, and, of course, all right angles are congruent. At first glance, this seems that this is a case of SSA, which is not sufficient to prove triangles congruent. Because the Pythagorean Theorem holds true for all right triangles, it is possible to show algebraically that the remaining sides are congruent as well.

A more geometric proof relies on isosceles triangles. In this proof, you place the two right triangles back-to-back, aligning them on the congruent legs.

Figure 4.36

You can be certain that the bottom edge is a line (because the two adjacent right angles form a straight angle), so you now have an isosceles triangle. In an isosceles triangle, base angles are congruent, and that additional bit of information lets you show the two right triangles congruent by AAS.

Example 1

If $\overline{OR} \perp \overline{KR}$, $\overline{OC} \perp \overline{KC}$, and $\overline{OR} \cong \overline{OC}$, prove $\overline{RK} \cong \overline{CK}$.

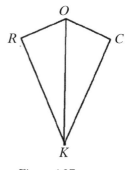

Figure 4.37

Solution: The given perpendiculars allow you to prove the triangles are right triangles. The shared segment $\overline{OK}$ is the hypotenuse of both triangles, and the given information provides the congruent legs. The triangles can be shown congruent by HL, and then CPCTC assures $\overline{RK} \cong \overline{CK}$.

Statement	Reason
1. $\overline{OR} \perp \overline{KR}$	1. Given
2. $\angle ORK$ is a right angle.	2. Perpendiculars form right angles.
3. $\triangle ORK$ is a right triangle. (◣)	3. A triangle that contains one right angle is a right triangle.
4. $\overline{OC} \perp \overline{KC}$	4. Given
5. $\angle OCK$ is a right angle.	5. Perpendiculars form right angles.
6. $\triangle OCK$ is a right triangle . (◣)	6. A triangle that contains one right angle is a right triangle.
7. $\overline{OK} \cong \overline{OK}$ (H)	7. Reflexive
8. $\overline{OR} \cong \overline{OC}$ (L)	8. Given
9. $\triangle ORK \cong \triangle OCK$	9. HL
10. $\overline{RK} \cong \overline{CK}$	10. CPCTC

Figure 4.38

Lesson 4-4 Review

1. For each pair of triangles, give the postulate or theorem that assures the triangles are congruent, or write "not congruent."

Figure 4.39

2. If $\overline{AE} \perp \overline{ML}$, $\overline{AM} \cong \overline{PL}$, and $\overline{ME} \cong \overline{PE}$, prove that $\triangle AEL$ is isosceles.

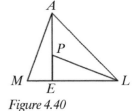

Figure 4.40

Lesson 4-5: Two Stage Congruence Proofs

Many times, you will encounter proofs that require you to prove one set of triangles congruent and then use CPCTC to get the information you need to prove a second set of triangles congruent. When you find that you don't have enough information to prove what you are asked to prove, ask yourself what you *can* prove. That may be the stepping stone to the information you need.

Example 1

If $\overline{CE} \cong \overline{CA}$, $\overline{ED} \cong \overline{AD}$, and $\overline{RD}$ bisects $\angle EDA$, prove $\triangle CER \cong \triangle CAR$.

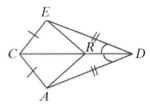

Figure 4.41

Solution: At first glance, it seems you have a lot of information in the hypothesis, but when you start to mark up the diagram, you realize that two of the three pieces of information don't apply to the triangles you are asked to prove congruent. The question to ask yourself then is: What can I prove?

There are two possibilities: either prove $\triangle DER \cong \triangle DAR$ (SAS), or prove $\triangle CED \cong \triangle CAD$ (SSS). Consider what help you will get from each one. If you prove $\triangle DER \cong \triangle DAR$, by CPCTC, you will have $\overline{ER} \cong \overline{AR}$, and you can prove $\triangle CER \cong \triangle CAR$ by SSS. If you prove $\triangle CED \cong \triangle CAD$, by CPCTC, you will have $\angle ECR \cong \angle ACR$, and you can prove $\triangle CER \cong \triangle CAR$ by ASA. Because either possibility will accomplish the goal, it is your choice which to do. The first possibility is shown in Figure 4.42.

Statement	Reason
1. $\overline{ED} \cong \overline{AD}$ (S $\triangle$1)	1. Given
2. $\overline{RD}$ bisects $\angle EDA$	2. Given
3. $\angle EDR \cong \angle ADR$ (A $\triangle$1)	3. A bisector divides an angle into two congruent angles.
4. $\overline{RD} \cong \overline{RD}$ (S $\triangle$1)	4. Reflexive
5. $\triangle DER \cong \triangle DAR$	5. SAS
6. $\overline{ER} \cong \overline{AR}$ (S $\triangle$2)	6. CPCTC
7. $\overline{CE} \cong \overline{CA}$ (S $\triangle$2)	7. Given
8. $\overline{CR} \cong \overline{CR}$ (S $\triangle$2)	8. Reflexive
9. $\triangle CER \cong \triangle CAR$	9. SSS

Figure 4.42

Example 2

Given: $\overline{ME} \cong \overline{XI}$, $\overline{EX} \cong \overline{MI}$, $\overline{EX} \parallel \overline{MI}$,

$\overline{MO} \perp \overline{EI}$ and $\overline{XC} \perp \overline{EI}$.

Prove: $\triangle MOE \cong \triangle XCI$

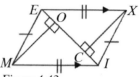

Figure 4.43

Solution: The given holds a lot of information. Mark it all on the diagram so that you
can see what moves you toward your goal. The perpendiculars tell
you that you have right triangles, and $\overline{ME} \cong \overline{XI}$ gives you the H of
hypotenuse-leg, but you don't immediately have the L. So turn your
attention to the other information and see what you can prove. The
parallel lines will let you say that alternate interior angles are congruent, so you could prove $\triangle MOI \cong \triangle XCE$ by AAS. CPCTC would
give you the leg you need.

Statement	Reason
1. $\overline{MO} \perp \overline{EI}$	1. Given
2. $\angle MOE$ and $\angle MOI$ are right angles.	2. Perpendiculars form right angles.
3. $\triangle MOE$ is a right triangle . (▲ △2)	3. A triangle that contains one right angle is a right triangle.
4. $\overline{XC} \perp \overline{EI}$	4. Given
5. $\angle XCI$ and $\angle XCE$ are right angles.	5. Perpendiculars form right angles.
6. $\triangle XCI$ is a right triangle . (▲ △2)	6. A triangle that contains one right angle is a right triangle.
7. $\angle MOI \cong \angle XCE$ (A △1)	7. All right angles are congruent.
8. $\overline{EX} \parallel \overline{MI}$	8. Given
9. $\angle XEI \cong \angle MIE$ (A △1)	9. If parallel lines are cut by a transversal, alternate interior angles are congruent.
10. $\overline{EX} \cong \overline{MI}$ (S △1)	10. Given
11. $\triangle MOI \cong \triangle XCE$	11. AAS
12. $\overline{XC} \cong \overline{MO}$ (L △2)	12. CPCTC
13. $\overline{ME} \cong \overline{XI}$ (H △2)	13. Given
14. $\triangle MOE \cong \triangle XCI$	14. HL

Figure 4.44

Lesson 4-5 Review

1. **Given:** $\overline{PR}$ and $\overline{AI}$ bisect one another at S.

   Prove: $\triangle PAR \cong \triangle RIP$

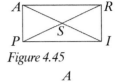

Figure 4.45

2. **Given:** $\triangle IOA$ is an isosceles triangle with
   $\overline{AI} \cong \overline{AO}$. M is the midpoint of $\overline{IO}$.
   $\angle LMI \cong \angle NMO$.

   Prove: $\triangle LMI \cong \triangle NMO$

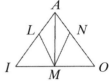

Figure 4.46

Answer Key

Lesson 4-1

1. $\triangle CAT \cong \triangle GOD$

2. $\angle ARM \cong \angle LEG$, $\angle RMA \cong \angle EGL$, $\angle MAR \cong \angle GLE$, $\overline{AR} \cong \overline{LE}$, $\overline{RM} \cong \overline{EG}$, and $\overline{AM} \cong \overline{LG}$.

Lesson 4-2

1. a. SAS   b. not congruent   c. SSS

2.

Statement	Reason
1. $\overline{WH} \perp \overline{WA}, \overline{EL} \perp \overline{LA}$	1. Given
2. $\angle HWA$ is a right angle, $\angle ELA$ is a right angle.	2. Perpendiculars form right angles.
3. $\angle HWA \cong \angle ELA$ (A)	3. All right angles are congruent.
4. $\overline{WA} \cong \overline{LA}$ (S)	4. Given
5. $\angle WAE \cong \angle LAH$	5. Given
6. $m\angle WAE = m\angle LAH$	6. Congruent angles have equal measures.
7. $m\angle HAE = m\angle HAE$	7. Reflexive
8. $m\angle WAE - m\angle HAE = m\angle LAH - m\angle HAE$	8. Subtraction Property of Equality
9. $m\angle WAE - m\angle HAE = m\angle WAH$, $m\angle LAH - m\angle HAE = m\angle LAE$	9. Angle Addition
10. $m\angle WAH = m\angle LAE$	10. Substitution
11. $\angle WAH \cong \angle LAE$ (A)	11. Congruent angles have equal measures.
12. $\triangle WHA \cong \triangle LEA$	12. ASA

Figure 4.47

3.

Statement	Reason
1. $\angle N \cong \angle H$ (A)	1. Given
2. $\angle ILO \cong \angle IOL$ (A)	2. Given
3. $\overline{DO} \cong \overline{LP}$	3. Given
4. $DO = LP$	4. Congruent segments have equal lengths.
5. $OL = OL$	5. Reflexive
6. $DO + OL = OL + LP$	6. Addition Property of Equality
7. $DO + OL = DL, OL + LP = OP$	7. Segment Addition
8. $DL = OP$	8. Substitution
9. $\overline{DL} = \overline{OP}$ (S)	9. Congruent segments have equal lengths.
10. $\triangle DLN \cong \triangle POH$	10. AAS

Figure 4.48

Lesson 4-3

1. $m\angle XYZ = 44°$

2. $m\angle YXZ = m\angle YZX$
    $5x + 3 = 111 - 4x$
    $9x = 108$
    $x = 12$
    $m\angle XYZ = 54°$ and $m\angle WXY = 117°$

3.

Statement	Reason
1. $\triangle LAK$ is a right triangle with right angle at A.	1. Given
2. $m\angle LAK = 90°$	2. A right angle measures 90°.
3. $m\angle LAK = m\angle LAE + m\angle KAE$	3. Angle Addition
4. $m\angle LAE + m\angle KAE = 90°$	4. Substitution
5. $\overline{LE} \cong \overline{AE} \cong \overline{EK}$	5. Given
6. $\angle L \cong \angle LAE$ and $\angle KAE \cong \angle K$	6. If two sides of a triangle are congruent, the angles opposite those sides are congruent.
7. $m\angle L = m\angle LAE$ and $m\angle KAE = m\angle K$	7. Congruent angles have equal measures.
8. $m\angle L + m\angle K = 90°$	8. Substitution
9. $\angle L$ and $\angle K$ are complementary.	9. If two angles add to 90° then they are complementary.

Figure 4.49

Lesson 4-4

1. a. HL   b. SSS   c. not congruent

2.

Statement	Reason
1. $\overline{AE} \perp \overline{ML}$	1. Given
2. $\angle MEA$ and $\angle PEL$ are right angles.	2. Perpendiculars form right angles.
3. $\triangle MEA$ and $\triangle PEL$ are right triangles. (▲)	3. A triangle that contains one right angle is a right triangle.
4. $\overline{AM} \cong \overline{PL}$ (H)	4. Given
5. $\overline{ME} \cong \overline{PE}$ (L)	5. Given
6. $\triangle MEA \cong \triangle PEL$	6. HL
7. $\overline{AE} \cong \overline{EL}$	7. CPCTC
8. $\triangle AEL$ is isosceles	8. A triangle with two congruent sides is an isosceles triangle.

Figure 4.50

Lesson 4-5

1.

Statement	Reason
1. $\overline{PR}$ and $\overline{AI}$ bisect one another at S.	1. Given
2. $\overline{PS} \cong \overline{RS}$ (S △1) and $\overline{AS} \cong \overline{IS}$ (S △1)	2. A bisector divides a segment into two congruent segments.
3. $\angle ASR \cong \angle PSI$ (A △1)	3. Vertical angles are congruent.
4. $\triangle PSI \cong \triangle RSA$	4. SAS
5. $\angle ARP \cong \angle IPR$ (A △2)	5. CPCTC
6. $\overline{AR} \cong \overline{IP}$ (S △2)	6. CPCTC
7. $\angle RAI \cong \angle PIA$ (A △2)	7. CPCTC
8. $\triangle PAR \cong \triangle RIP$	8. ASA

Figure 4.51

2.

Statement	Reason
1. $\triangle IOA$ is an isosceles triangle with $\overline{AI} \cong \overline{AO}$.	1. Given
2. $\angle I \cong \angle O$ (A $\triangle$1)	2. If two sides of a triangle are congruent, the sides opposite them are congruent.
3. M is the midpoint of $\overline{IO}$	3. Given
4. $\overline{IM} \cong \overline{OM}$ (S $\triangle$1)	4. A midpoint divides the segment into two congruent segments.
5. $\angle LMI \cong \angle NMO$ (A $\triangle$1)	5. Given
6. $\triangle LMI \cong \triangle NMO$	6. ASA
7. $\overline{LM} \cong \overline{NM}$ (S $\triangle$2)	7. CPCTC
8. $\overline{IL} \cong \overline{ON}$	8. CPCTC
9. $AI = AO$ and $IL = ON$	9. Congruent segments have equal measures.
10. $AI = AL + IL$	10. Segment Addition
11. $AO = AN + ON$	11. Segment Addition
12. $AL + IL = AN + ON$	12. Substitution
13. $AL = AN$	13. Subtraction Property of Equality
14. $\overline{AL} \cong \overline{AN}$ (S $\triangle$2)	14. Congruent segments have equal length.
15. $\overline{AM} \cong \overline{AM}$ (S $\triangle$2)	15. Reflexive
16. $\triangle LMI \cong \triangle NMO$	16. SSS

Figure 4.52

Inequalities

Lesson 5-1: Properties of Inequality

The properties of inequality are covered in algebra, and most look very similar to the properties of equality. The most important exception is the multiplication property. The **multiplication property of equality** says that if you multiply both sides of an equation by the same number, the equation remains in balance. When you multiply both sides of an inequality by a number, however, you need to be very conscious of the sign of the multiplier. When you multiply both sides of that inequality by a negative number, the relationship changes. It may be true that $7 > 4$, but $-3 \cdot 7 < -3 \cdot 4$. Remember that the negative side of the number line is a mirror image of the positive side. Multiplying both sides of an inequality by a negative number reverses the direction of the inequality.

The other caution when working with inequalities is transitivity. There is a transitive property for inequality, but when you use it, you need to be alert to the direction of the inequality. The **transitive property of inequality** says that if $a > b$ and $b > c$, then $a > c$. The property could be restated using $<$ instead of $>$: if $a < b$ and $b < c$, then $a < c$. You must be certain that you have a linkage with the inequality signs in the same direction. If $a > b$ and $c > b$, you can't apply the transitive property, because you don't have the chain, and a rewrite won't help, because $c > b$ becomes $b < c$, and that means the inequality signs don't match. Use the transitive property of inequality carefully.

Example 1

Determine if each of the statements is true or false.

a. If $m\angle A = m\angle B$ and $m\angle B + m\angle C > m\angle D$,
   then $m\angle A + m\angle C > m\angle D$.

b. If $x < y$ and $y > z -7$, then $x < z - 7$.

c. If $a + c > b + c$ and $d > 0$ then $d(a + c) > d(b + c)$.

Solution:

a. True. Because $m\angle A$ is equal to $m\angle B$, it is legal to substitute one for the other, even in an inequality.

b. False. The transitive property does not apply, because the inequality signs do not point in the same direction.

c. True. Multiplying both sides of an inequality by the same positive number leaves the inequality in the same direction.

Lesson 5-1 Review

Determine if each of the statements is true or false.

1. If $AB > BC$ and $BC > CD + DE$, then $AB > CD + DE$.

2. If $p < q$ and $p + r > t$, then $q + r > t$.

3. If $m\angle R < m\angle T$ and $m\angle T < m\angle W$, then $m\angle R < m\angle W$.

4. If $x < 0$ and $y > z$, then $xy > xz$.

5. If $ST < TU$ and $RS + ST = RT$, then $RS + TU < RT$.

Lesson 5-2: The Exterior Angle Theorem

An **exterior angle** of a triangle is an angle formed by extending one side of triangle. If side $\overline{AC}$ of $\triangle ABC$ is extended through C to a point D, it forms $\angle BCD$. If side $\overline{AB}$ is extended through B to E, $\angle CBE$ is formed. Both angles ($\angle BCD$ and $\angle CBE$) are exterior angles of $\triangle ABC$.

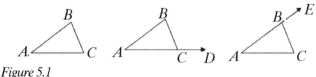

Figure 5.1

An exterior angle and the interior angle adjacent to it form a linear pair, and therefore they are supplementary. In the center triangle in Figure 5.1, exterior angle $\angle BCD$ and interior angle $\angle ACB$ are a linear pair, so m$\angle ACB$ + m$\angle BCD$ = 180°.

The three angles in any triangle total 180°, so m$\angle ACB$ + m$\angle CBA$ + m$\angle BAC$ = 180°. Because m$\angle ACB$ + m$\angle BCD$ = 180°, m$\angle ACB$ + m$\angle CBA$ + m$\angle BAC$ = m$\angle ACB$ + m$\angle BCD$. Subtracting m$\angle ACB$ from both sides gives you m$\angle CBA$ + m$\angle BAC$ = m$\angle BCD$. The measure of an exterior angle of a triangle is equal to the sum of the measures of the two interior angles that are not adjacent to it. Those angles are the **remote interior angles**.

Example 1

In $\triangle ABC$, side $\overline{AB}$ is extended through B to E to form exterior angle $\angle CBE$. If m$\angle A = 43°$ and m$\angle C = 61°$, find the measure of $\angle CBE$.

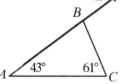

Figure 5.2

Solution: The three angles of a triangle total 180°, so you know that m$\angle A$ + m$\angle ABC$ + m$\angle C$ = 180°. Substituting, you get 43 + m$\angle ABC$ + 61 = 180, or m$\angle ABC$ + 104 = 180, which tells you that m$\angle ABC$ = 76°. Because $\angle CBE$ is supplementary to $\angle ABC$, $\angle CBE$ must measure 104°.

The faster method of solving this problem is just to remember that the measure of an exterior angle of a triangle is equal to the sum of the measures of the remote interior angles. Then you can simply add 43 + 61 to find that m$\angle CBE$ = 104°.

According to a fundamental postulate of inequality, the whole is greater than its part. Because m$\angle CBE$ = m$\angle A$ + m$\angle C$, the whole (m$\angle CBE$) is greater than m$\angle A$ or m$\angle C$. In general, *the measure of an exterior angle of a triangle is greater than either of the remote interior angles.*

Example 2

Label each statement true or false.

a. m$\angle ANE$ > m$\angle NGE$

b. m$\angle EGL$ = m$\angle AEG$

c. m$\angle ENG$ < m$\angle EAN$

d. m$\angle NGE$ > m$\angle GEL$

e. m$\angle ENG$ + m$\angle NEG$ = m$\angle GAE$ + m$\angle AEG$

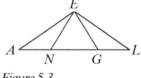

Figure 5.3

Solution:

a. True. ∠*ANE* is an exterior angle of Δ*NGE*, so m∠*ANE* is greater than either of the remote interior angles. ∠*NGE* is a remote interior angle, so m∠*ANE* > m∠*NGE*.

b. False. ∠*EGL* is an exterior angle of Δ*AEG,* and so it is greater than the remote interior angle ∠*AEG*.

c. False. ∠*ENG* is an exterior angle of Δ*AEN*, so it is larger than the remote interior angle ∠*EAN*.

d. True. ∠*NGE* is an exterior angle of Δ*GEL*, so it is larger than remote interior angle ∠*GEL*.

e. True. Focus on Δ*ENG*, which contains ∠*ENG* and ∠*NEG*. It has an exterior angle (∠*EGL*) so m∠*EGL* = m∠*ENG* + m∠*NEG*. Because ∠*EGL* is also an exterior angle of Δ*AEG*, m∠*EGL* = m∠*GAE* + m∠*AEG*.
By transitivity, m∠*ENG* + m∠*NEG* = m∠*GAE* + m∠*AEG*.

Lesson 5-2 Review

Use Δ*POE* with exterior angle ∠*HOP* to answer the following questions.

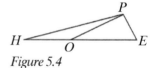

Figure 5.4

1. If m∠*OPE* = 85° and m∠*E* = 50°, find the measure of ∠*HOP*.

2. If m∠*HOP* = 137° and m∠*OPE* = 91°, find the measure of ∠*E*.

3. If m∠*OPE* = 12*x* + 5, m∠*E* = 8*x* – 2, and m∠*HOP* = 18*x* + 17, find *x*.

4. True or False: m∠*POE* > m∠*H*

5. True or False: m∠*HPO* > m∠*POE*.

Lesson 5-3: Inequalities in Triangles

The exterior angle inequality relates an angle outside the triangle to an angle inside. Some of the most useful inequalities, however, make connections between sides and angles, either in one triangle, or between two triangles.

One of the most helpful inequalities says that the largest side of a triangle lies opposite the largest angle, and the shortest side is opposite the smallest angle. If ∠*Z* is the largest angle in Δ*XYZ*, then $\overline{XY}$ is the longest side of the triangle. If $\overline{XZ}$ is the shortest side, then ∠*Y* is the smallest angle.

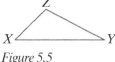
Figure 5.5

Proving that relationship means proving both:

If *XY* > *YZ* > *XZ*, then m∠*1* > m∠*2* > m∠*3*.

and

If m∠*1* > m∠*2* > m∠*3*, then *XY* > *YZ* > *XZ*.

To prove the first statement, extend $\overline{XZ}$, and find a point *W* on $\overline{XZ}$, so that $\overline{YW}$ is the same length

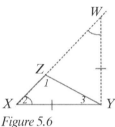
Figure 5.6

as $\overline{XY}$. Because $\overline{XY} \cong \overline{XW}$, the opposite angles are congruent, so ∠*2* ≅ ∠*W*. ∠*1* is an exterior angle of Δ*ZWY*, so m∠*1* > m∠*W*. Because ∠*2* ≅ ∠*W*, you can substitute to show m∠*1* > m∠*2*. By repeating this type of argument with a different extended side, you can show that m∠*2* > m∠*3*, and then, by transitivity, m∠*1* > m∠*2* > m∠*3*.

To prove the converse, an indirect proof is the easiest method. You want to prove that, if m∠*1* > m∠*2* > m∠*3*, then *XY* > *YZ* > *XZ*. An indirect proof will try to prove the contrapositive, which says that if *XY* ≤ *YZ* ≤ *XZ*, then m∠*1* ≤ m∠*2* ≤ m∠*3*. If *XY* ≤ *YZ* ≤ *XZ*, then the inequality can be written as *XZ* ≥ *YZ* ≥ *XY*. If *XZ* ≥ *YZ* ≥ *XY*, then, according to the theorem just proven, m∠*3* ≥ m∠*2* ≥ m∠*1*. If m∠*3* ≥ m∠*2* ≥ m∠*1*, then rewrite as m∠*1* ≤ m∠*2* ≤ m∠*3*. That proves the contrapositive, so the conditional is also true. Therefore, if m∠*1* > m∠*2* > m∠*3*, then *XY* > *YZ* > *XZ*.

In simple language, the largest angle is opposite the longest side.

Example 1

Complete each statement with the logical conclusion from the hypothesis given.

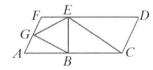
Figure 5.7

a. If m∠*BEC* > m∠*BCE*, then

_____ .

b. If *ED* > *EC*, then _____ .

c. If *GE* = *GB*, then _____ .

d. If m∠*EGB* > m∠*GEB*, then _____ .

e. If *GE* > *FE*, then _____ .

Solution:

a. If m∠*BEC* > m∠*BCE*, then *BC* > *EB*. In Δ*BEC*, $\overline{BC}$ is opposite the larger angle, ∠*BEC*, so $\overline{BC}$ is the longer side.

b. If *ED* > *EC*, then m∠*ECD* > m∠*D*. ∠*ECD* is opposite the longer side, $\overline{ED}$, so ∠*ECD* is the larger angle.

c. If *GE* = *GB*, then m∠*GBE* = m∠*GEB*, because base angles of an isosceles triangle are congruent.

d. If m∠*EGB* > m∠*GEB*, then *EB* > *GB*, because in Δ*EGB*, the larger side is opposite the larger angle.

e. If *GE* > *FE*, then m∠*F* > m∠*FGE*, because in Δ*FGE*, the larger angle is opposite the longer side.

So far, the inequalities have talked about a single triangle, connecting an exterior angle to the remote interior angles or relating sides and angles within the triangle. Another theorem, often referred to as the **Hinge Theorem**, allows you to write inequalities that connect two triangles. The hypothesis of the theorem sounds similar to a congruent triangle theorem. Unlike congruence statements, however, the Hinge Theorem requires that the angles included between two congruent sides be different, not congruent. If you have those conditions (congruent corresponding sides and non-congruent included angles), then the triangle with the larger included angle has the larger third side.

The name Hinge Theorem comes from the visual image of the two triangles. Think about a door, opening and closing on its hinges. The lengths of the sides don't change, but the angle between them does.

Figure 5.8

What the Hinge Theorem tells you is that the larger that angle becomes, the longer the segment you will need to complete the triangle.

Figure 5.9

Applied to two different triangles, this means that, if $AB = XY$, $AC = XZ$, and if $m\angle A < m\angle X$, then $BC < YZ$. Because

Figure 5.10

$\overline{AB}$ is the same size as $\overline{XY}$, and $\overline{AC}$ is the same size as $\overline{XZ}$, the difference between the triangles is how far the hinge is open. The hinge is open wider in $\triangle XYZ$, so $\overline{YZ}$ is longer than $\overline{BC}$.

You can prove this theorem by placing $\triangle ABC$ and $\triangle XYZ$ on top of one another with the shorter corresponding sides ($\overline{AB}$ and $\overline{XY}$) aligned.

If you draw $\overline{CZ}$, you will create an isosceles triangle ($\triangle ZXC$) because $AC = XZ$, and its base angles are congruent, so $\angle XZC \cong \angle XCZ$. Next, notice that $\angle ZCY$ is larger than $\angle ZCX$ and $\angle XZC$ is larger than $\angle YZC$.

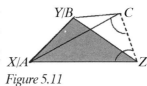

Figure 5.11

You have three relationships:

1. $\angle XZC \cong \angle ZCX$

2. $m\angle ZCY > m\angle ZCX$

3. $m\angle XZC > m\angle YZC$

If $m\angle ZCY > m\angle XCZ$ (2) and $m\angle XCZ = m\angle XZC$ (1), then $m\angle ZCY > m\angle XZC$.
If $m\angle ZCY > m\angle XZC$ and $m\angle XZC > m\angle YZC$ (3), then $m\angle ZCY > m\angle YZC$.
If $m\angle ZCY > m\angle YZC$,
then in $\triangle YCZ$, $YZ > YC$, because the longer side is opposite the larger angle. But $YC = BC$, so $YZ > BC$.

Example 2

If $\overline{HE} \cong \overline{HI}$, prove that $m\angle IEG > m\angle EIG$.

Solution: This is a case in which the plan is probably best built from the bottom up. To prove that $m\angle IEG > m\angle EIG$, you probably want to find the triangle that contains both angles ($\triangle IEG$) and show that $\overline{IG}$ is longer than $\overline{EG}$. Taking a step back from there, you can use the

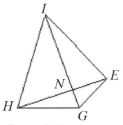

Figure 5.12

Hinge Theorem to show that $IG > EG$, if you can show corresponding sides of two triangles congruent and the included angles different. Use ΔHGE and ΔHGI for this. $\overline{HG}$ is in both triangles, and the given provides the other pair of congruent sides.

Statement	Reason
1. $\overline{HE} \cong \overline{HI}$	1. Given
2. $\overline{HG} \cong \overline{HG}$	2. Reflexive
3. $m\angle IHG > m\angle EHG$	3. The whole is greater than its part.
4. $IG > EG$	4. Hinge Theorem
5. $m\angle IEG > m\angle EIG$	5. In a triangle (ΔIEG), the larger angle is opposite the longer side.

Figure 5.13

Example 3

$EF = FO$, $FO \perp OD$, and $m\angle DFO > m\angle EFD$.
Which is larger: $\angle DEO$ or $\angle DOE$?

Solution: Look at ΔFED and ΔFOD. Because $EF = FO$, $FD = FD$, and $m\angle DFO > m\angle EFD$, $DO > DE$. Focusing on ΔDEO, you see that $DO > DE$ means that $m\angle DEO > m\angle DOE$.

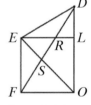

Figure 5.14

Lesson 5-3 Review

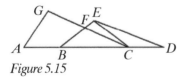

Figure 5.15

1. Complete each statement with the logical conclusion from the hypothesis given.

a. If $m\angle G > m\angle GAC$, then _____.

b. If $ED > EB$, then _____.

c. If $AG = BE$, $AC = BD$, and $m\angle GAC > m\angle EBD$, then _____.

2. $AC = BD$, $BE = EC$, $BC = ED$, and m∠BCE > m∠CED. If $AB = 5$, find the smallest possible value of BE.

3. If $BC > EC$ and △BEC is isosceles, prove that m∠BEC > m∠EBC.

Lesson 5-4: The Triangle Inequality Theorem

The **Triangle Inequality Theorem** deals with a single triangle, but talks only about sides, not angles. You've used the theorem, almost intuitively, even if you haven't given it an official name. If you need to get from point A to point B, you know, without a formal explanation, that it is shorter to walk through the park, on a straight line from A to B, than to walk from A to C and then C to B.

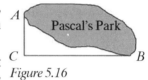

Figure 5.16

The Triangle Inequality Theorem says that the length of the third side of a triangle is less than the sum of the other two sides. In that statement, the "third side" could be any side of the triangle, so you're actually creating three inequalities. In symbols, you would say that in △ABC:

$$AB < BC + CA$$
$$BC < AB + CA$$
$$CA < AB + BC$$

If you subtract CA from both sides of the middle inequality, you find that $BC - CA < AB$. Combining the inequalities gives you $BC - CA < AB < BC + CA$. The length of any side of a triangle is less than the sum of the other two sides and greater than their difference.

Example 1

In △RST, $RS = 12$ and $ST = 8$. Find the largest and the smallest possible values for RT.

Solution: $RS - ST < RT < RS + ST$, so $12 - 8 < RT < 12 + 8$. RT must be greater than 4 and less than 20.

The Triangle Inequality Theorem often gives you a wide range of values as possibilities for the third side. When the angle between the two known sides is large, the length of the third side will be closer to the sum.

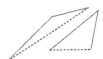

Figure 5.17

When the angle is small, the third side is closer to the difference between the lengths of the known sides.

Example 2

The distance from New York to Los Angeles is 2,451 miles. The distance from Chicago to Los Angeles is 1,744 miles. Find the largest and smallest possible values for the distance from New York to Chicago. Which end of the range is the better estimate?

Figure 5.18

Solution: The distance from New York to Chicago is greater than 2,451 − 1,744 and less than 2,451 + 1,744 miles. The smallest possible distance is 707 miles. The largest possible distance is 4,195 miles. One look at a map tells you that Chicago and New York are much closer than Los Angeles and New York, so the upper limit is too large. Because the side of the triangle that represents the distance from New York to Chicago lies opposite one of the very small angles in the triangle, the smaller value is probably the better estimate. In fact, the distance between New York and Chicago is 714 miles.

Lesson 5-4 Review

1. In $\triangle ABC$, $AB = 3$ and $AC = 7$. Find the largest and the smallest possible values for BC.

2. In $\triangle XYZ$, $XY = 23$ and $YZ = 8$. Which of these could NOT be the length of XZ?
   a. 14   c. 21
   b. 18   d. 30

3. In $\triangle RST$, $RS = 3x - 9$, $ST = 5x + 2$, and $RT = 9x - 17$. Find the largest and the smallest possible values for the positive integer x.

4. The distance from Caracas, Venezuela, to Madrid, Spain, is 4,351 miles. The distance from Madrid to Hong Kong is 6,555 miles. Find the largest and smallest possible values for the distance from Caracas to Hong Kong.

Answer Key

Lesson 5-1

1. True
2. True
3. True
4. False
5. False

Lesson 5-2

1. $m\angle HOP = 135$
2. $m\angle E = 46°$
3. $x = 7$
4. True
5. False

Lesson 5-3

1. a. $AC > GC$
   b. $m\angle EBD > m\angle EDB$
   c. $GC > ED.$
2. $BE > 5$
3.

Statement	Reason
1. $BC > EC$ $\triangle BEC$ is isosceles	1. Given
2. $BE = EC$	2. Definition of isosceles
3. $BC > BE$	3. Substitution
4. $m\angle BEC > m\angle EBC$	4. The larger angle is opposite the longer side.

Figure 5.19

Lesson 5-4

1. $4 < BC < 10$
2. Choice **a** cannot be the length of $\overline{XZ}$ because $15 < XZ < 31$, so XZ cannot equal 14.
3. $4 < x < 10.$
4. The distance from Caracas to Hong Kong is greater than 2,204 miles and less than 10,906 miles.

Similarity

Lesson 6-1: Ratio and Proportion

A **ratio** is a comparison of two numbers—specifically, a comparison by division. The most common way to write a ratio is a fraction. In fact, the name for numbers that can be written as fractions is rational numbers, and the word *rational*, in that context, comes from ratio.

Two equal ratios form a proportion. $\frac{18}{27} = \frac{2}{3}$ is an example of a proportion. When all four numbers are shown, the proportion is simply a statement of two equal fractions. When one or more of the numbers in the proportion are unknown, the proportion becomes an equation to be solved.

The principal technique for solving a proportion is an application of the means-extremes property. That name comes from the fact that in a proportion the four positions for numbers have labels. Two of them are means, and two are extremes. The pattern of the labels is $\frac{\text{extreme}}{\text{mean}} = \frac{\text{mean}}{\text{extreme}}$ or extreme: mean = mean: extreme. In the proportion $\frac{18}{27} = \frac{2}{3}$, 18 and 3 are the extremes and 27 and 2 are the means.

> The *means-extremes property* says that the product of the means is equal to the product of the extremes.

In $\frac{18}{27} = \frac{2}{3}$, the product of the means is $2 \cdot 27$, and the product of the extremes is $18 \cdot 3$. Both products are 54. In any proportion, these two products will be the same. This gives you a tactic most simply called cross-multiplying, which can be used to solve a proportion.

In the proportion $\frac{5}{3} = \frac{8}{x}$, one number is unknown, but the product of the means will still equal the product of the extremes. That means $5x = 3 \cdot 8$, which can easily be solved to find that $x = 4.8$.

Example 1

Solve for x: $\dfrac{27}{8} = \dfrac{x}{24}$

Solution: Cross-multiply to produce the equation

$$8x = 27 \cdot 24 \Rightarrow 8x = 648 \Rightarrow x = \frac{648}{8} = 81.$$

Example 2

Solve for x: $\dfrac{3}{5} = \dfrac{x+1}{x-1}$

Solution: Cross-multiply to produce the equation $3(x - 1) = 5(x + 1)$. Distribute to get $3x - 3 = 5x + 5$ and solve.

$$3x - 3 = 5x + 5$$
$$-2x - 3 = 5$$
$$-2x = 8$$
$$x = -4$$

Although proportions are usually simple to solve, cross-multiplying can occasionally produce quadratic equations. You will probably find that these can be solved by factoring, but you can resort to the quadratic formula if you need it.

Example 3

Solve for x: $\dfrac{2x-3}{5} = \dfrac{3}{x+5}$

Solution: Cross-multiplying gives you $(2x - 3)(x + 5) = 5 \cdot 3$, or $2x^2 + 7x - 15 = 15$. Subtract 15 from both sides to get $2x^2 + 7x - 30 = 0$. Factor to get $(2x - 5)(x + 6) = 0$. Setting each factor equal to zero and solving gives you $2x - 5 = 0 \Rightarrow 2x = 5 \Rightarrow x = \dfrac{5}{2}$

and $x + 6 = 0 \Rightarrow x = -6$.

If you look back at your algebra or pre-algebra book, you'll probably find a list of properties of proportions. One important property of proportions talks about exchanging the means or the extremes. It says that if $\frac{a}{b} = \frac{c}{d}$, then $\frac{a}{c} = \frac{b}{d}$ or $\frac{d}{b} = \frac{c}{a}$. When you cross-multiply $\frac{a}{b} = \frac{c}{d}$, you get $ad = bc$. Divide both sides by cd and you'll have $\frac{a}{c} = \frac{b}{d}$. If instead you divide both sides by ab, you get $\frac{d}{b} = \frac{c}{a}$. Exchange both the means and the extremes, and you'll have another important property of proportions: If $\frac{a}{b} = \frac{c}{d}$, then $\frac{d}{c} = \frac{b}{a}$. Stated in different terms: If two ratios are equal, their reciprocals are equal as well.

The other property that will be useful requires some examination. It says that if $\frac{a}{b} = \frac{c}{d}$, then $\frac{a \pm b}{b} = \frac{c \pm d}{d}$. If $\frac{a}{b} = \frac{c}{d}$, then by cross-multiplication $ad = bc$. Add bd to both sides and you have $ad + bd = bc + bd$. Now factor, and that equation becomes $d(a + b) = b(c + d)$. Finally divide both sides

by bd: $\dfrac{\cancel{d}(a+b)}{b\cancel{d}} = \dfrac{\cancel{b}(c+d)}{\cancel{b}d} \Rightarrow \dfrac{a+b}{b} = \dfrac{c+d}{d}$.

Lesson 6-1 Review

Solve for the unknown in each proportion.

1. $\dfrac{3}{x} = \dfrac{15}{35}$

3. $\dfrac{2x - 1}{3x} = \dfrac{5}{9}$

5. $\dfrac{x+1}{x-3} = \dfrac{x+3}{5}$

2. $\dfrac{x+4}{2} = \dfrac{x-5}{7}$

4. $\dfrac{4}{x} = \dfrac{x}{16}$

Answer True or False for each question.

6. If $\dfrac{AB}{XY} = \dfrac{BC}{YZ}$, then $\dfrac{YZ}{XY} = \dfrac{BC}{AB}$.

7. If $\dfrac{AB}{XY} = \dfrac{BC}{YZ}$, then $\dfrac{AB}{XY} = \dfrac{XZ}{BC}$.

8. If $\dfrac{AB}{XY} = \dfrac{BC}{YZ}$, then $\dfrac{XY}{AB} = \dfrac{YZ}{BC}$.

9. If $\dfrac{AB}{BC} = \dfrac{XY}{YZ}$, then $\dfrac{AB + BC}{BC} = \dfrac{XY + YZ}{YZ}$.

10. If $\dfrac{AB}{BC} = \dfrac{XY}{YZ}$, then $\dfrac{AB}{BC - AB} = \dfrac{XY}{YZ - XY}$.

Lesson 6-2: Similar Figures

Imagine a projector throwing an image on a screen. The distance between the projector and the screen determines the size of the image. Moving the projector closer to the screen makes the image smaller; pulling the projector away makes the image larger. Changing the size of the image in that way wouldn't be very useful if it distorted the image, however. Whether you make the image larger or smaller, you would want it to have the same shape.

When a polygon is enlarged or reduced, the size must change, but the shape must remain. Size is determined by the length of sides, so the lengths of corresponding sides change. In order to maintain the shape, however, all sides must change in the same proportion, and the angles must remain the same size.

Polygons that are the same shape but different sizes are called similar polygons. **Similar polygons** have corresponding angles congruent, but corresponding sides are not congruent. The ratio of corresponding sides is always the same, however.

> **Two polygons are similar if and only if their corresponding angles are congruent and their corresponding sides are in proportion.**

When you write a congruence statement, the order in which you name the vertices of the figures tells your reader the correspondence, and the same is true of similarity statements. If you write $\triangle ABC \sim \triangle RST$, you are telling your reader that $\angle A \cong \angle R$, $\angle B \cong \angle S$, and $\angle C \cong \angle T$. The relationship of the sides follows from that correspondence. The ratio of AB to RS is the same as the ratio of BC to ST and the ratio of AC to RT, so

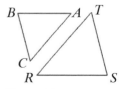

Figure 6.1

you can write the extended proportion $\frac{AB}{RS} = \frac{BC}{ST} = \frac{AC}{RT}$. The ratio of corresponding sides is called the **ratio of similitude** or the scale factor.

Example 1

Determine which of the triangles are similar, write the similarity statement, and give the ratio of similitude.

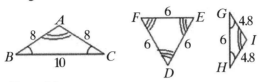

Figure 6.2

Solution: Two of the triangles are isosceles and one is an equilateral triangle. Because there is no way that the angles of an isosceles triangle could be congruent to the angles of an equilateral triangle, you can eliminate $\triangle DEF$. You cannot assume that the remaining triangles are similar, however. You must check the markings on the triangles to be certain that corresponding angles are congruent and that corresponding sides are in proportion. According to those markings, $\angle B \cong \angle C \cong \angle G \cong \angle H$, and $\angle A \cong \angle I$. The ratio of BC to GH is $\frac{10}{6} = \frac{5}{3}$, and if the ratio of AB:GI or AC:HI is simplified $\frac{8}{4.8} = \frac{80}{48} = \frac{5}{3}$. Corresponding angles are congruent and corresponding sides are in proportion, so the two isosceles triangles are similar. This correspondence can be written either as $\triangle ABC \sim \triangle IGH$ or $\triangle ABC \sim \triangle IHG$.

The ratio of similitude (or scale factor) is $\frac{5}{3}$.

Example 2

If two quadrilaterals are similar, with $\square BLUE \sim \square PINK$, list the conclusions that can be drawn.

Solution: Because the quadrilaterals are similar, corresponding angles are congruent, and the congruence statement tells which angles correspond. $\angle B \cong \angle P$, $\angle L \cong \angle I$, $\angle U \cong \angle N$, and $\angle E \cong \angle K$.

Corresponding sides are in proportion, so $\frac{BL}{PI} = \frac{LU}{IN} = \frac{UE}{NK} = \frac{BE}{PK}$.

If you know that two polygons are similar, you can use the fact that corresponding sides are in proportion to find missing lengths. You must know the ratio of similitude or the lengths of a pair of corresponding sides, from which you can calculate the ratio. Once you have that, you can set up a proportion and solve the proportion to find the length of a side.

Example 3

$\square BLUE \sim \square PINK$. Find the length of $\overline{EU}$.

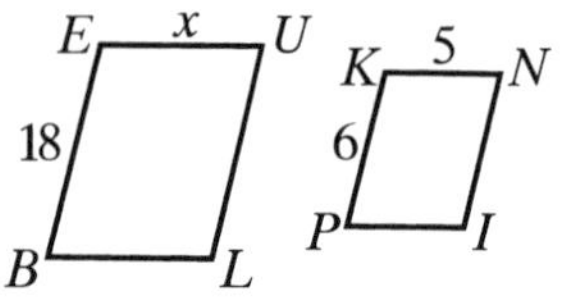

Solution: The polygons are similar, so corresponding sides are in proportion, and the congruence statement $\square BLUE \sim \square PINK$ tells you which sides correspond.

Figure 6.3

The known sides are $\overline{BE}$, $\overline{PK}$, and $\overline{KN}$, so you want a proportion involving those and $\overline{EU}$. $\dfrac{BE}{PK} = \dfrac{EU}{KN}$ becomes $\dfrac{18}{6} = \dfrac{x}{5}$ when you substitute the known values. Cross-multiply and solve to get $90 = 6x \Rightarrow x = 15$.

Example 4

If $\triangle RST \sim \triangle XYZ$, $RS = 4$, $RT = 3$, $XY = 3t - 2$ and $XZ = 2t + 5$, find the value of t and the lengths of $\overline{XY}$ and $\overline{XZ}$.

Solution: Because $\triangle RST \sim \triangle XYZ$, corresponding sides are in proportion. $\dfrac{RS}{XY} = \dfrac{RT}{XZ}$ becomes $\dfrac{4}{3t-2} = \dfrac{3}{2t+5}$ when the given information is substituted.

Cross-multiplying produces $4(2t + 5) = 3(3t - 2)$, and solving gives you $8t + 20 = 9t - 6$ or $t = 26$. Because $t = 26$, $3t - 2 = 3(26) - 2 = 76$, and $2t + 5 = 2(26) + 5 = 57$. Therefore $XY = 76$ and $XZ = 57$.

Lesson 6-2 Review

If the polygons are similar, write the similarity statement and give the ratio of similitude. If they are not similar, explain why.

1.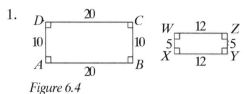

Figure 6.4

2.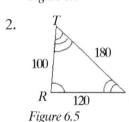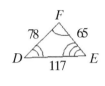

Figure 6.5

3. If $\triangle RED \sim \triangle TAN$, list the conclusions that can be drawn.

4. In Figure 6.6, $\triangle FIR \sim \triangle ELM$. Find the length of $\overline{EL}$.

5. In Figure 6.7, $FOST \sim ORES$. If $FT = x - 1$, $OS = x$, and $RE = 4$, find the lengths of $\overline{FT}$ and $\overline{OS}$.

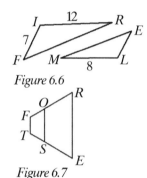

Figure 6.6

Figure 6.7

Lesson 6-3: Proving Triangles Similar

When you learned to prove triangles congruent, you saw that there were shortcuts. In the same way, there are some shortcuts for proving triangles similar. When you are trying to show that two triangles are similar, it is enough to prove that two pairs of corresponding angles are congruent. You do not have to show that the third pair of angles is congruent, or that corresponding sides are in proportion. You know that the three angles in any triangle add up to 180 degrees, so, if two angles in one triangle are congruent to the corresponding angles in the other triangle, a little arithmetic will prove that the third pair of angles will be congruent as well.

Once it's been shown that proving two pairs of corresponding angles congruent is enough to guarantee that the triangles are similar, you can use that shortcut in other proofs. You will find that you use it so often that it qualifies for our list of subroutines.

Subroutine

Proving Triangles Similar

$\angle A \cong \angle X$ (A)	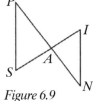
$\angle B \cong \angle Y$ (A)	
$\triangle ABC \sim \triangle XYZ$ AA	

Figure 6.8

Example 1

Given: $\overline{PS} \parallel \overline{IN}$

Prove: $\triangle PSA \sim \triangle NIA$

Solution: Use alternate interior angles to show that two pair of corresponding angles are congruent.

Figure 6.9

Statement	Reason
1. $\overline{PS} \parallel \overline{IN}$	1. Given
2. $\angle P \cong \angle N$ (A)	2. If parallel lines are cut by a transversal, alternate interior angles are congruent.
3. $\angle S \cong \angle I$ (A)	3. If parallel lines are cut by a transversal, alternate interior angles are congruent.
4. $\triangle PSA \sim \triangle NIA$	4. AA

Figure 6.10

Example 2

Given: $\overline{EP} \cong \overline{ER}$, $\angle S \cong \angle U$

Prove: $\dfrac{SR}{PU} = \dfrac{SE}{UC}$

Figure 6.11

Solution: The only method you have to prove that the lengths of segments are in proportion is to prove that they are the corresponding sides of similar polygons. Looking at the proportion, you see that $\overline{SR}$ and $\overline{SE}$ are sides of $\triangle SRE$ and that $\overline{PU}$ and $\overline{UC}$ are sides of $\triangle UPC$. You want to prove $\triangle SRE \sim \triangle UPC$, using AA.

Statement	Reason
1. $\angle S \cong \angle U$ (A)	1. Given
2. $\overline{EP} \cong \overline{ER}$	2. Given
3. $\triangle PER$ is an isosceles triangle.	3. If two sides of a triangle are congruent, the triangle is isosceles. (Definition of isosceles triangle)
4. $\angle ERS \cong \angle UPC$ (A)	4. Base angles of an isosceles triangle are congruent.
5. $\triangle SRE \sim \triangle UPC$	5. AA
6. $\dfrac{SR}{PU} = \dfrac{SE}{UC}$	6. Corresponding sides of similar triangles are in proportion.

Figure 6.12

Lesson 6-3 Review

1. **Given:** $\overline{AL} \parallel \overline{ME}$

   Prove: $\triangle PAL \sim \triangle MPE$

2. **Given:** $\angle PIE \cong \angle N$, and $\angle E \cong \angle IPN$

   Prove: $\dfrac{PI}{PE} = \dfrac{IN}{PI}$

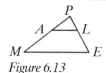

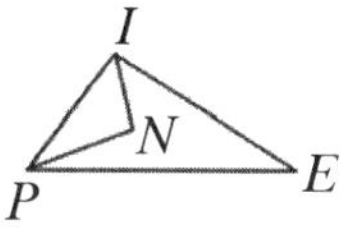

Figure 6.13

Figure 6.14

Lesson 6-4: Side-Splitting

When you think about similar triangles, you imagine two triangles, but those triangles might be arranged in such a way that one seems to be part of the other. In such a case, you may be proving parts of sides proportional to whole sides. For example, if you are given that $\overline{AL} \parallel \overline{ME}$, you can prove $\triangle PAL \sim \triangle MPE$ by AA, and then conclude that $\frac{PA}{PM} = \frac{PL}{PE}$, which is $\frac{\text{part}}{\text{whole}} = \frac{\text{part}}{\text{whole}}$.

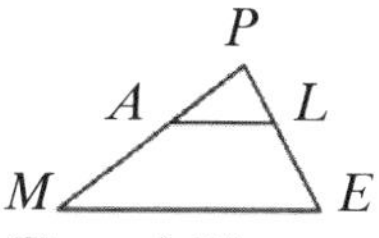

Figure 6.15

With this result, one of the properties of proportions, and a little work, you can prove the **Side Splitter Theorem**. It says that a line parallel to one side of a triangle divides the other two sides proportionally. Rather than $\frac{\text{part}}{\text{whole}} = \frac{\text{part}}{\text{whole}}$, the Side Splitter Theorem says:

$$\frac{\text{top part}}{\text{bottom part}} = \frac{\text{top part}}{\text{bottom part}}$$

Assume you're given $\triangle ABC$, and line $\overleftrightarrow{DE} \parallel \overline{AC}$. $\overrightarrow{DE}$ splits side $\overline{BC}$ into $\overline{BE}$ and $\overline{EC}$, and side $\overline{BA}$ into $\overline{BD}$ and $\overline{DA}$. You want to show that

Figure 6.16

$\frac{BE}{EC} = \frac{BD}{DA}$. If $\overleftrightarrow{ED}$ is parallel to $\overleftrightarrow{AC}$, then $\angle BED \cong$ $\angle BCA$ and $\angle BDE \cong \angle BAC$, because, when parallel lines are cut by a transversal, corresponding angles are congruent. $\triangle BED \sim \triangle BCA$ by AA, and $\frac{BE}{BC} = \frac{BD}{BA}$, because corresponding sides of similar triangles are in proportion. However, $BC = BE + EC$ and $BA =$ $BD + DA$, by segment addition, and so you can substitute to produce

$\frac{BE}{BE + EC} = \frac{BD}{BD + DA}$. Apply properties of proportions to prove $\frac{BE}{EC} = \frac{BD}{DA}$.

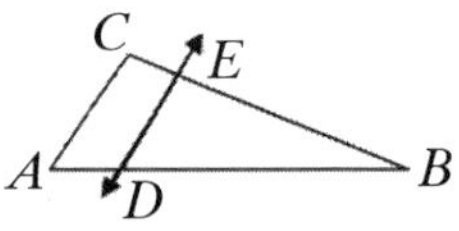

The converse is also true. If a line divides two sides of a triangle proportionally, then it is parallel to the third side. You're given $\triangle ABC$ and line $\overleftrightarrow{DE}$, which divides the sides so that $\frac{BE}{EC} = \frac{BD}{DA}$. It's possible to prove $\triangle BED \sim \triangle BCA$, and so $\angle BED \cong \angle BCA$ and $\angle BDE \cong \angle BAC$. Finally, because corresponding angles are congruent, the lines are parallel.

Example 1

Given: $\overline{CU} \parallel \overline{IR} \parallel \overline{AT}$

Prove: $AI \cdot RU = IC \cdot TR$

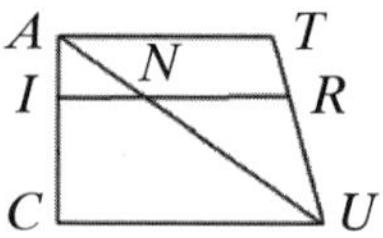

Figure 6.17

Solution: The equal products look strange, but they came from cross-multiplying a proportion. Think backwards to see what the proportion might have been.

$\dfrac{AI}{IC} = \dfrac{TR}{RU}$ is one possibility.

If you use the Side Splitter Theorem in $\triangle CAU$, and then use it again in $\triangle TUA$, the two proportions should connect and let you use the transitive property to produce that proportion.

Statement	Reason
1. $\overline{CU} \parallel \overline{IR} \parallel \overline{AT}$	1. Given
2. $\dfrac{AI}{IC} = \dfrac{AN}{NU}$	2. Side Splitter Theorem
3. $\dfrac{AN}{NU} = \dfrac{TR}{RU}$	3. Side Splitter Theorem
4. $\dfrac{AI}{IC} = \dfrac{TR}{RU}$	4. Transitive Property
5. $AI \cdot RU = IC \cdot TR$	5. Means-Extremes Property (Cross-multiplying)

Figure 6.18

Example 2

In Figure 6.17, $AN = x + 3$, $AU = 15$, $AI = x + 1$, and $AC = 12$, find the value of x.

Solution: According to the Side Splitter Theorem, $\overline{UP} \parallel \overline{TI}$. The given information doesn't quite fit that proportion, so you have a choice: either use the properties of proportions to find a proportion that fits your given information, or do a bit of arithmetic to find the lengths IC and NU. The latter sounds easier. $IC = AC - AI = 12 - (x + 1) = 11 - x$.
$NU = AU - AN = 15 - (x + 3) = 12 - x$.

Then $\overline{UT} \parallel \overline{PC}$ becomes $\overline{UE} \cong \overline{TP}$. Cross-multiplying produces an equation that looks to be a quadratic for a moment, but simplifies quickly.

$$(x+1)(12-x) = (11-x)(x+3)$$
$$12x - x^2 + 12 - x = 11x + 33 - x^2 - 3x$$
$$-x^2 + 11x + 12 = -x^2 + 8x + 33$$
$$11x + 12 = 8x + 33$$
$$3x + 12 = 33$$
$$3x = 21$$
$$x = 7$$

Related to the Side Splitter Theorem is one that talks about the midsegment of a triangle. The **midsegment** is a line segment that joins the midpoints of two sides of a triangle. The theorem says that the midsegment of a triangle is parallel to the third side and half as long.

Given $\triangle XYZ$ with M the midpoint of side $\overline{XZ}$ and N the midpoint of side $\overline{YZ}$, prove that $\overline{MN} \parallel \overline{XY}$ and $MN = \frac{1}{2}XY$. Because M and N are the midpoints of $\overline{XZ}$ and $\overline{YZ}$, $XM = MZ$ and $YN = NZ$. That means that $\frac{XM}{MZ} = \frac{YN}{NZ} = 1$. The converse of the side splitter theorem tells you that $\overline{MN} \parallel \overline{XY}$. Once you know the segments are parallel, you can prove $\triangle MNZ \sim \triangle XYZ$ by AA. Corresponding sides of similar triangles are in proportion, so $\frac{MN}{XY} = \frac{NZ}{YZ} = \frac{MZ}{XZ} = \frac{1}{2}$. Focus on $\frac{MN}{XY} = \frac{1}{2}$, and multiply both sides by XY to prove $MN = \frac{1}{2}XY$.

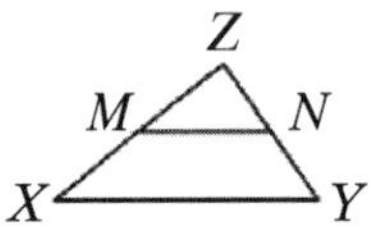

Figure 6.19

Example 3

If L is the midpoint of $\overline{BI}$ and N is the midpoint of $\overline{ID}$, find the lengths of $\overline{LN}$ and $\overline{BD}$.

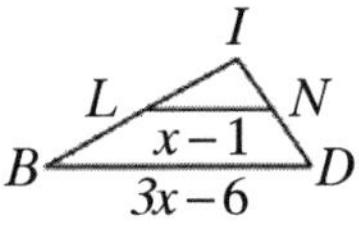

Figure 6.20

Solution: By the midsegment theorem, $\overline{LN} \parallel \overline{BD}$ and half as long, so

$$x - 1 = \tfrac{1}{2}(3x - 6)$$
$$2(x - 1) = 3x - 6$$
$$2x - 2 = 3x - 6$$
$$-2 = x - 6$$
$$4 = x$$

Therefore, $LN = x - 1 = 3$ and $BD = 3(4) - 6 = 6$.

The next theorem isn't technically a side-splitting theorem, but it talks about how parallel lines divide their transversals. Imagine you have three parallel lines and two transversals cutting through them. The theorem says that the parallels divide the transversals proportionally.

As soon as you hear the word *proportionally*, you should be thinking about similar triangles, but the given information doesn't seem to include any triangles. Make your own

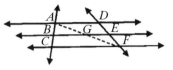

triangles, by drawing $\overline{AF}$. In $\triangle ACF$, the Side   *Figure 6.21*

Splitter Theorem tells you that $\frac{AB}{BC} = \frac{AG}{GF}$. Use the same theorem on $\triangle DFA$

to show that $\frac{DE}{EF} = \frac{AG}{GF}$. The transitive property tells you that $\frac{AB}{BC} = \frac{DE}{EF}$.

Example 4

If $RD = x + 4$, $RA = x$, $PE = 9$, and $PS = 21$, find the

length of $\overline{DA}$.

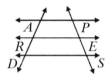

Solution: The parallel lines divide the

transversals proportionally, so $\dfrac{RA}{RD} = \dfrac{PE}{ES}$, and all   *Figure 6.22*

but one of these values appears in the given information. You need to find ES, by subtracting $PS - PE = 21 - 9 = 12$. Then:

$$\frac{x}{x+4} = \frac{9}{12}$$
$$12x = 9(x+4)$$
$$12x = 9x + 36$$
$$3x = 36$$
$$x = 12$$
$$DA = RD + RA = x + 4 + x = 12 + 4 + 12 = 28$$

The final side-splitting theorem talks about what happens when you bisect one angle of a triangle. An **angle bisector** in a triangle divides the opposite side into two segments that are proportional to the sides adjacent to them. In terms of Figure 6.23, that means that if $\overrightarrow{ZW}$ bisects

$\angle XZY$, then $\frac{XW}{WY} = \frac{XZ}{YZ}$. You might occasionally come

across a triangle in which an angle bisector creates two similar triangles, but this illustration is clearly a counterexample. $\triangle XWZ$ is definitely not similar to $\triangle YWZ$. So where is the proportion coming from?

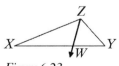

Figure 6.23

The trick is to draw a line through X, parallel to $\overrightarrow{ZW}$, and extend $\overline{YZ}$ until they meet at V. Focus on $\triangle XYV$, and the Side Splitter Theorem tells you $\frac{YZ}{ZV} = \frac{WY}{XW}$. Three of the segments in that proportion are part of the proportion you're trying to prove, but ZV is not. What can you find out about ZV?

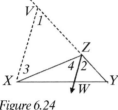

Figure 6.24

You drew $\overline{XV} \parallel \overrightarrow{ZW}$, so the corresponding angles and alternate interior angles are congruent. That means $\angle 1 \cong \angle 2$ and $\angle 3 \cong \angle 4$, but because $\overrightarrow{ZW}$ is the bisector of $\angle XZY$, $\angle 2 \cong \angle 4$. Use transitivity and you have $\angle 1 \cong \angle 2 \cong \angle 4 \cong \angle 3$, so $\angle 1 \cong \angle 3$. If two angles of a triangle are congruent, the sides opposite those angles are congruent, so $\overline{ZV} \cong \overline{XZ}$. Substitute in $\frac{YZ}{ZV} = \frac{WY}{XW}$ and you have $\frac{YZ}{XZ} = \frac{WY}{XW}$. Apply the property of proportions that says that if two ratios are equal, their reciprocals are equal, and you have $\frac{XW}{WY} = \frac{XZ}{YZ}$.

Example 5

In Figure 6.25, if $\overline{AB}$ bisects $\angle JAM$, find the length of $\overline{JM}$.

Solution: If $\overline{AB}$ bisects $\angle JAM$, then
$$\frac{JB}{JA} = \frac{BM}{AM}, \text{ so } \frac{x}{10} = \frac{2x-3}{15}.$$
Cross-multiplying, $15x = 10(2x - 3)$, and $15x = 20x - 30$, so $x = 6$.
Therefore, $JM = x + 2x - 3 = 6 + 12 - 3 = 15$.

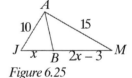

Figure 6.25

Lesson 6-4 Review

1. If $\overline{XY} \parallel \overline{AC}$, $BY = 3$, $YC = x - 1$, $AX = x + 7$, and $XB = 5$ in Figure 6.26, find the length of $\overline{AB}$.

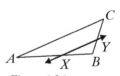

Figure 6.26

2. Jim's house is at the midpoint of the road from Pascoag to Chepatchet and Jon's house is at the midpoint of the road from Pascoag to Putnam. If Putnam is 14 miles from Chepatchet, how far is it from Jon's house to Jim's house?

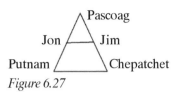

Figure 6.27

3. In Figure 6.28, if *AR* = *t* + 1, *RD* = 6, *PE* = 4, and *ES* = *t* − 1, find the length of $\overline{PS}$.

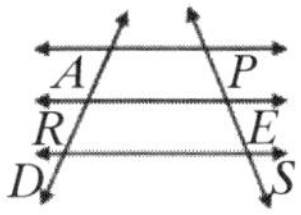

Figure 6.28

4. If Δ*ABC* is a right triangle with right angle at ∠*ABC*, and if $\overline{BD}$ bisects ∠*ABC*, prove *AD* · *BC* = *AB* · *CD*.

Lesson 6-5: Perimeters and Areas of Similar Figures

Because congruent polygons are exact duplicates of one another, they have the same dimensions, the same perimeter, and the same area. Similar polygons, on the other hand, are different sizes, so their perimeters and areas will not be the same. Because the lengths of corresponding sides are in proportion, there should be relationships between the perimeters and between their areas.

Rectangles are probably the easiest to start with, because the perimeter and area of rectangles are so easy to calculate. Suppose □*ABCD* ~ □*WXYZ*, with the dimensions shown in Figure 6.29. The ratio of similitude is $\frac{12}{9.6} = \frac{5}{4}$.

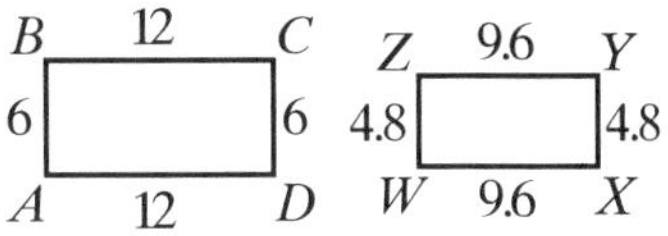

Figure 6.29

The perimeter of □*ABCD* is 2(12) + 2(6) = 24 + 12 = 36. The perimeter of □*WXYZ* is 2(9.6) + 2(4.8) = 19.2 + 9.6 = 28.8. The ratio of the perimeters is $\frac{36}{28.8} = \frac{5}{4}$, exactly the same as the ratio of similitude.

The area of □*ABCD* is 6(12) or 72 square units. The area of □*WXYZ* is 4.8(9.6) = 46.08. The ratio of the areas is $\frac{72}{46.08} = \frac{25}{16}$. This ratio doesn't match the ratio of similitude; instead, it is the square of that scale factor.

The **perimeter** of any polygon is found by simply adding up the lengths of its sides. If you have a pair of similar polygons, and you know the lengths of the sides of one polygon, you can find the lengths of the other by multiplying the known sides by the scale factor. If, for example, you had similar pentagons, and scale factor is k, the sides of one pentagon could be represented by s_1, s_2, s_3, s_4, and s_5, and the sides of the other by ks_1, ks_2, ks_3, ks_4, and ks_5. The perimeter of the first pentagon is $P = s_1 + s_2 + s_3 + s_4 + s_5$ and the perimeter of the second is $P = ks_1 + ks_2 + ks_3 + ks_4 + ks_5$.

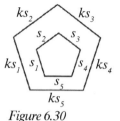

Figure 6.30

This second perimeter can be rewritten as:

$P = k (s_1 + s_2 + s_3 + s_4 + s_5)$, or exactly k times the perimeter of the first polygon.

Example 1

EARTH and *VENUS* are pentagons and *EARTH ~VENUS*.
If $AR = 18$, $EN = 6$, and the perimeter of *VENUS* is 42, what is the perimeter of *EARTH*?

Solution: $\overline{AR}$ and $\overline{EN}$ are corresponding sides of similar

polygons, so the ratio of similitude or scale factor is $\dfrac{AR}{EN} = \dfrac{18}{6} = 3$.
The ratio of the perimeters is that same, so

$\dfrac{\text{perimeter of } EARTH}{\text{perimeter of } VENUS} = \dfrac{x}{42} = 3$. Multiply both sides by 42 to find
that the perimeter of *EARTH* is 126.

Before you start to investigate areas of similar polygons, it's important to note that any polygon can be broken down into triangles. Pick one vertex and draw diagonals from that corner to every vertex you can.

Figure 6.31

As you start to explore areas of similar polygons, you don't need to worry about all the different possible polygons. If two polygons are similar, and you break both of them into triangles, starting from corresponding vertices, the triangles that are created will be similar. The area of the polygon can be found by adding the areas of the triangles. So the rest of the investigation on area of similar polygons is going to look just at similar triangles.

Figure 6.32

In order to calculate the area of a triangle, you need to know the length of a side that will serve as the base, and the length of an altitude or height, drawn from the opposite vertex perpendicular to the base. If two triangles are similar, their sides are in proportion, and, with a little effort, you can prove that the heights are in proportion as well.

If $\triangle ABC \sim \triangle RST$, then $\angle A \cong \angle R$. Because $\overline{CD} \perp \overline{AB}$ and $\overline{TV} \perp \overline{RS}$, $\angle ADC$ and $\angle RVT$ are right angles, and $\angle ADC \cong \angle RVT$, because all right angles are congruent. Therefore $\triangle ADC \sim \triangle RVT$, by AA, and

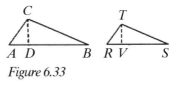

Figure 6.33

$\frac{AC}{RT} = \frac{CD}{TV}$. The ratio of the corresponding heights is the same as the ratio of the corresponding sides.

Example 2

$\triangle SAT \sim \triangle URN$. The altitude from A perpendicular to $\overline{ST}$ measures 28 and the altitude from R perpendicular to $\overline{UN}$ measures 7. If the altitude from S to $\overline{AT}$ is 34 units, find the length of the altitude from U to $\overline{RN}$.

Solution: In similar triangles, the ratio of the corresponding heights is equal to the ratio of corresponding sides, so

$\frac{\text{altitude from } A}{\text{altitude from } R} = \frac{28}{7} = 4$, and the scale factor is 4. The ratio of the altitude from S to the altitude from U must also be 4, so

$\frac{\text{altitude from } S}{\text{altitude from } U} = \frac{34}{x} = 4$.

Cross-multiplying, you get $4x = 34$ or $x = 8.5$. The length of the altitude from U to $\overline{RN}$ is 8.5 units.

The area of a triangle is calculated by the formula $A = \frac{1}{2}bh$, where b is the length of the base and h is the length of the altitude drawn perpendicular to that base. When two triangles are similar, the base and height of one triangle, multiplied by the scale factor, give you the corresponding base and height of the second triangle. If b and h are the base and height of one triangle, and k is the scale factor, the base and height of the other triangle are kb and kh. The area of the first triangle is $A = \frac{1}{2}bh$

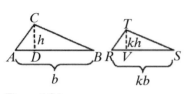

Figure 6.34

and the area of the second is $A = \frac{1}{2}(kb)(kh) = k^2\left(\frac{1}{2}bh\right)$. The ratio of the areas is k^2, the square of the scale factor.

Example 3

If $\triangle MET \sim \triangle EOR$, and the area of $\triangle EOR$ is 7.5 square units, find the area of $\triangle MET$.

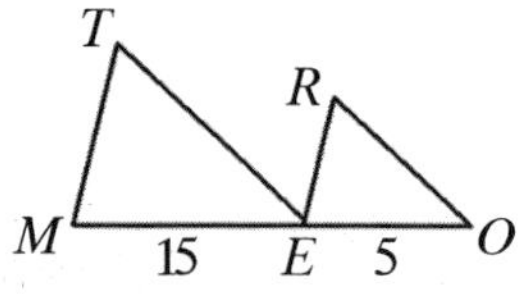

Solution: The ratio of the corresponding

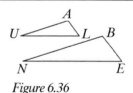

Figure 6.35

sides $\dfrac{ME}{EO} = \dfrac{15}{5} = 3$, so the ratio of the areas

will be $3^2 = 9$. Therefore, $\dfrac{\text{Area of } \triangle MET}{\text{Area of } \triangle EOR} = \dfrac{x}{7.5} = 3^2 = 9$, and solving $x = 7.5(9) = 67.5$. The area of $\triangle MET$ is 67.5 square units.

Example 4

$\triangle NEB \sim \triangle ULA$. The area of $\triangle NEB$ is 175 square units, and the area of $\triangle ULA$ is 63 square units. $BE = 7 - x$ and $AL = x + 3$. Find the length of $\overline{AL}$.

Figure 6.36

Solution: $\dfrac{\text{Area of } \triangle NEB}{\text{Area of } \triangle ULA} = \dfrac{175}{63} = \dfrac{25}{9}$. The ratio of the areas is the

square of the ratio of the sides, so if $\left(\dfrac{BE}{AL}\right)^2 = \dfrac{25}{9}$, then $\dfrac{BE}{AL} = \dfrac{5}{3}$.

With that information, you can write the proportion $\dfrac{7-x}{x+3} = \dfrac{5}{3}$.

Cross-multiplying produces

$$3(7-x) = 5(x+3)$$
$$21 - 3x = 5x + 15$$
$$6 = 8x$$
$$x = \dfrac{3}{4}$$

The length of $\overline{AL}$ is $\dfrac{3}{4} + 3 = 3\dfrac{3}{4}$ units.

Lesson 6-5 Review

1. In trapezoid $ABCD$, $BC = 27$ and total perimeter is 149. In trapezoid $WXYZ$, $XY = 51$. Find the total perimeter of $WXYZ$.

2. $\triangle RST \sim \triangle DEF$. Altitude $\overline{SV}$ measures 30 inches and altitude $\overline{EG}$ measures 18 inches. If $RS = 45$ inches, find DE.

3. If $\triangle ABC \sim \triangle XYZ$, $BC = 19$ units, $YZ = 66.5$ units, and the area of $\triangle ABC$ is 75 square units, find the area of $\triangle XYZ$.

4. $\triangle DOG \sim \triangle CAT$. The area of $\triangle DOG$ is 4 square units, and the area of $\triangle CAT$ is 81 square units. $DG = 4x + 1$ and $CT = 27 - 2x$. Find the length of $\overline{DG}$.

Answer Key

Lesson 6-1

1.
$$\dfrac{3}{x} = \dfrac{15}{35}$$
$$15x = 105$$
$$x = \dfrac{105}{15} = 7$$

2.
$$\frac{x+4}{2} = \frac{x-5}{7}$$
$$7(x+4) = 2(x-5)$$
$$7x+28 = 2x-10$$
$$5x = -38$$
$$x = -7.6$$

3.
$$\frac{2x-1}{3x} = \frac{5}{9}$$
$$9(2x-1) = 5 \cdot 3x$$
$$18x-9 = 15x$$
$$-9 = -3x$$
$$x = 3$$

4.
$$\frac{4}{x} = \frac{x}{16}$$
$$x^2 = 64$$
$$x = \pm 8$$

5.
$$\frac{x+1}{x-3} = \frac{x+3}{5}$$
$$5(x+1) = (x-3)(x+3)$$
$$5x+5 = x^2 - 9$$
$$0 = x^2 - 5x - 14$$
$$0 = (x-7)(x+2)$$
$$x-7 = 0 \Rightarrow x = 7$$
$$x+2 = 0 \Rightarrow x = -2$$

6. True   7. False   8. True   9. True   10. True

Lesson 6-2

1. $\square ABCD$ and $\square WXYZ$ are not similar because the sides are not in proportion.

2. $\triangle RST \sim \triangle EDF$. Ratio of similitude: $\dfrac{20}{13}$

3. If $\triangle RED \sim \triangle TAN$, $\angle R \cong \angle T$, $\angle E \cong \angle A$, $\angle D \cong \angle N$ and $\dfrac{RE}{TA} = \dfrac{ED}{AN} = \dfrac{RD}{TN}$.

4. $$\dfrac{FI}{IR} = \dfrac{EL}{LM}$$

$$\dfrac{7}{12} = \dfrac{x}{8}$$

$$12\,x = 56$$

$$x = \dfrac{56}{12} = \dfrac{14}{3} = 4\dfrac{2}{3}$$

5. Set up a proportion using the fact that $FOST \sim ORES$. The ratio of FT (from $FOST$) to OS (from $ORES$) is equal to the ratio of OS (from $FOST$) to RE (from $ORES$), so $\dfrac{FT}{OS} = \dfrac{OS}{RE}$.

$$\dfrac{FT}{OS} = \dfrac{OS}{RE}$$

$$\dfrac{x-1}{x} = \dfrac{x}{4}$$

$$4(x-1) = x^2$$

$$4x - 4 = x^2$$

$$x^2 - 4x + 4 = 0$$

$$(x-2)^2 = 0$$

$$x - 2 = 0$$

$$x = 2$$

Therefore, $FT = x - 1 = 2 - 1 = 1$ and $OS = x = 2$.

Lesson 6-3

1.

Statement	Reason
1. $\overline{AL} \parallel \overline{ME}$	1. Given
2. $\angle PAL \cong \angle PME$ (A)	2. If parallel lines are cut by a transversal, corresponding angles are congruent.
3. $\angle PLA \cong \angle PEM$ (A)	3. If parallel lines are cut by a transversal, corresponding angles are congruent.
4. $\triangle PAL \sim \triangle MPE$	4. AA

Figure 6.37

2.

Statement	Reason
1. $\angle PIE \cong \angle N$ (A)	1. Given
2. $\angle E \cong \angle IPN$ (A)	2. Given
3. $\triangle PIE \sim \triangle INP$	3. AA
4. $\dfrac{PI}{IN} = \dfrac{PE}{PI}$	4. Corresponding sides of similar triangles are in proportion.
5. $\dfrac{PI}{PE} = \dfrac{IN}{PI}$	5. Property of Proportions

Figure 6.38

Lesson 6-4

1.
$$\frac{BY}{YC} = \frac{BX}{XA}$$
$$\frac{3}{x-1} = \frac{5}{x+7}$$
$$3(x+7) = 5(x-1)$$
$$3x + 21 = 5x - 5$$
$$26 = 2x$$
$$x = 13$$
$$AB = AX + XB = 13 + 7 + 5 = 25$$

2. The road from Jon's house to Jim's house connects the midpoints of two sides of the triangle, so it is a midsegment of the triangle. That means that the road connecting their houses, parallel to the road from Putnam to Chepatchet, is 14 miles, so the distance from Jon's house to Jim's house is half of 14, or 7, miles.

3.
$$\frac{AR}{RD} = \frac{PE}{ES}$$
$$\frac{t+1}{6} = \frac{4}{t-1}$$
$$(t+1)(t-1) = 24$$
$$t^2 - 1 = 24$$
$$t^2 = 25$$
$$t = \pm 5$$
$$PS = PE + ES = 4 + 5 - 1 = 8$$

4.

Statement	Reason
1. $\overline{BD}$ bisects $\angle ABC$	1. Given
2. $\dfrac{AD}{DC} = \dfrac{AB}{BC}$	2. In a triangle, the bisector of an angle divides the opposite side into two segments proportional to the adjacent sides.
3. $\dfrac{AD}{AB} = \dfrac{DC}{BC}$	3. Property of Proportions
4. $AD \cdot BC = AB \cdot CD$	4. Means-Extremes Property (Cross-multiplication)

Figure 6.39

Lesson 6-5

1. $\dfrac{BC}{XY} = \dfrac{\text{perimeter of } ABCD}{\text{perimeter of } WXYZ} \Rightarrow \dfrac{27}{51} = \dfrac{149}{x} \Rightarrow \dfrac{9}{17} = \dfrac{149}{x}$.

$9x = 2533$

$x = 281\dfrac{4}{9}$

The perimeter of $WXYZ$ is $281\dfrac{4}{9}$.

2. $\dfrac{SV}{EG} = \dfrac{RS}{DE} \Rightarrow \dfrac{30}{18} = \dfrac{45}{x} \Rightarrow x = 27$

$DE = 27$

3.   The ratio of similitude is $\dfrac{BC}{YZ} = \dfrac{19}{66.5} = \dfrac{2}{7}$.

The ratio of the areas is equal to the square of the ratio of similitude.

$\dfrac{75}{x} = \left(\dfrac{2}{7}\right)^2$

$\dfrac{75}{x} = \dfrac{4}{49}$

$4x = 3675$

$x = 918.75$

The area of $\triangle XYZ$ is 918.75 square units.

4. The ratio of the areas is $\dfrac{4}{81}$, which is $\left(\dfrac{2}{9}\right)^2$, so the ratio of the sides is $\dfrac{2}{9}$

$$\frac{DG}{CT} = \frac{2}{9}$$

$$\frac{4x+1}{27-2x} = \frac{2}{9}$$

$$9(4x+1) = 2(27-2x)$$

$$36x + 9 = 54 - 4x$$

$$40x = 45$$

$$x = \frac{45}{40} = \frac{9}{8}$$

$$DG = 4\left(\frac{9}{8}\right) + 1 = \frac{9}{2} + \frac{2}{2} = \frac{11}{2} = 5.5$$

Transformations

Somewhere in your study of algebra, you encounter the notion of a function. Sometimes a function is portrayed as a machine: You put a number in, the function machine works on it, and gives you something back. The *squaring* function takes in a 5 and gives back a 25. The *add four* function takes in −1 and gives back 3.

In geometry, where you're dealing with points, lines, and polygons instead of numbers, that function concept has to be a little different, but the idea still exists. The term **transformation** refers to a group of actions that can work on a figure to produce a new figure. The transformation sends the points of the original figure to new points, according to a rule. The original figure is the preimage, and the result of the transformation is the image. If the preimage is a point P, the image will often be designated as P'. If the preimage is a polygon, for example, $\square ABCD$, its image under a transformation might be designated as $\square A'B'C'D'$.

In this chapter, you'll learn about four kinds of transformations: reflections, translations, rotations, and dilations. You'll see that the first three are rigid transformations, meaning that they do not change the size or shape of the figure, but simply move it or create a copy of it in a new position. Rigid transformations create images congruent to the preimage, and they are called isometries. Dilations, on the other hand, will change the size of the figure, but they will increase or decrease the size proportionally, so the image will be similar to the original figure.

In algebra, functions can be combined in various ways to form new functions. Transformations can be combined as well, one following another. The first transformation works on the preimage—say $\triangle ABC$—and produces an image, $\triangle A'B'C'$. The next transformation works on that image, producing $\Delta A''B''C''$, and so on. Some common combinations of transformations produce effects that get names of their own, such as the glide reflection, which is a combination of a translation (or glide) and a reflection.

Lesson 7-1: Reflections

Everyone has some experience with reflections. If you checked yourself out in a mirror when you got dressed this morning, you took for granted that your reflection was a reasonably accurate copy of how you actually look. If you stop to think about that reflection, you realize that there are crucial differences, of which the most obvious is the left/right reversal.

Geometry talks about the reflection of a point over a line, and the reflecting line acts as the surface of the mirror. To find the reflection of a point P over a line l, draw line m through P, perpendicular to l. Think of

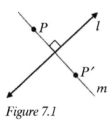

Figure 7.1

m as a ray of light from P hitting the mirror. Where does the image of P appear to be when you look at the mirror? It seems to be on the other side of the mirror's surface at a distance equal to the distance from P to the mirror surface. The image of point P, under the reflection R_l, is a point P' on m, positioned so that line l is the perpendicular bisector of $\overline{PP'}$.

When you are working in the coordinate plane, it's common to see reflections over one of the axes, or over the line $y = x$, but any line can serve as a reflecting line. A reflection over the x-axis might be abbreviated as Rx and a reflection over the y-axis as Ry. Reflections over other lines would be shown as an upper case R with the equation of the line noted as a subscript (for example, $R_{y\,=\,x}$).

Example 1

Find the image of the point $P(3, -2)$ after a reflection:

a. over the x-axis   b. over the y-axis   c. over the line $y = x$

Solution:

a. The x-axis is horizontal so the line through P perpendicular to the x-axis is the vertical line $x = 3$. P is 2 units below the x-axis, so the image point, P', will be on $x = 3$, 2 units above the x-axis. P' is the point $(3, 2)$.

b. The y-axis is vertical, so the line perpendicular to it through P is the horizontal line $y = -2$. P is 3 units right of the axis, so P' is on the line $y = -2$ and 3 units left of the axis. P' is the point $(-3, -2)$.

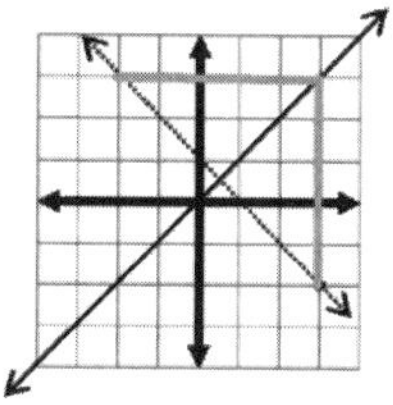

c. For a reflection over the line $y = x$, you need a line perpendicular to $y = x$, so remember that perpendicular lines have slopes that are negative reciprocals. The slope of $y = x$ is 1, so the slope of any line perpendicular to it is -1. You want the perpendicular to pass through P so it must be

Figure 7.2

the line $y - (-2) = -1(x - 3)$ or $y = -x + 1$. Rather than doing a long calculation with the distance formula to determine exactly which point on that line is the image point, use what you know about isosceles triangles. If you draw a vertical line connecting P to the line $y = x$ at $(3, 3)$ and then connect that point to $y = -x + 1$ at $(-2, 3)$ with a horizontal line, you create an isosceles triangle. The line $y = x$ is the altitude to the base of that triangle so it bisects the base. The point $(-2, 3)$ is on the line $y = -x + 1$, and is as far from $y = x$ as P is, so P' is the point $(-2, 3)$.

That may seem to be a lot of work just to find the reflection of one point, but if you take note of the patterns, you can skip the work in the future.

Preimage	Reflected Over	Image	Shortcut
$(3, -2)$	The x-axis	$(3, 2)$	Change the sign of the y-coordinate
$(3, -2)$	The y-axis	$(-3, -2)$	Change the sign of the x-coordinate
$(3, -2)$	The line $y = x$	$(-2, 3)$	Exchange the x- and y-coordinates

Figure 7.3

Example 2

Find the image of each point under the specified reflection.

a. $(3, -9)$ reflected over the y-axis

b. $(-6, 1)$ reflected over the x-axis

c. $(4, 0)$ reflected over $y = x$

d. (0, –5) reflected over the x-axis

e. (0, –5) reflected over the y-axis

f. (0, –5) reflected over y = x

Solution:

a. Change the sign of the x-coordinate: the image is (–3, –9)

b. Change the sign of the y-coordinate: the image is (–6, –1)

c. Exchange the x- and y- coordinates: the image is (0, 4)

d. Change the sign of the y-coordinate: the image is (0, 5)

e. Change the sign of the x-coordinate: the image is (0, –5). Because (0, –5) is on the reflecting line, the y-axis, it doesn't appear to move. The preimage and the image are the same point.

f. Exchange the x- and y-coordinates: the image is (–5, 0)

To find the reflection image of a polygon, in theory, you need to reflect every point in the polygon. Because that would be infinitely many points, it's not a practical strategy. Instead, reflect each of the vertices of the polygon, and connect them to form the image polygon.

Example 3

$\triangle ABC$ has vertices $A(-2, 1)$, $B(6, 2)$, and $C(4, -1)$. Draw the reflection of $\triangle ABC$ over the line $y = x$.

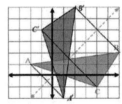

Solution: Using the shortcut for reflecting over $y = x$, you can determine that A' is (1, –2), B' is (2, 6), and C' is (–1, 4). Plot these points and connect A' to B' to C' and back to A'.

Figure 7.4

As you connect the vertices to form the image triangle you may notice that you're moving counterclockwise, whereas in the preimage, A to B to C and back to A is a clockwise motion. Reflections reverse orientation, just as left and right are reversed when you look in a mirror. Notice, though, that lengths and angle measurements are not disturbed. The image and the preimage are congruent. Reflection preserves distance, length, and angle measure.

Lesson 7-1 Review

1. Draw the image of each polygon under the reflection specified.

  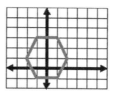

a. Over the *x*-axis   b. Over the *y*-axis   c. Over the line *y* = *x*

Figure 7.5

Lesson 7-2: Translations

If a polygon is reflected over a line, the image is a polygon congruent to the preimage, but with a different orientation. If the image is reflected again, over a line parallel to the first one, the second image created is congruent to the other polygons, and has the same orientation as the preimage. If you erase the intermediate image and look only at the preimage and the final image, it will look as though the polygon slid to its new location.

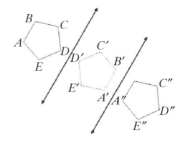

Figure 7.6

The translation, or slide, is actually the composition of two reflections, over parallel lines, but it appears to be one of the simplest transformations. It takes the preimage figure and moves it, point for point, according to a specific rule about distance and direction. If you take a pentagon and slide it half an inch to the right, that's a translation. When an elevator moves from floor to floor, the passenger car is translated up and down.

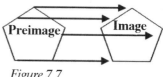

Figure 7.7

The movement doesn't have to be restricted to vertical or horizontal; oblique translation is common. Translation can be described by giving the image and the two parallel reflecting lines, but the simpler method is to give the direction and the distance of the movement, the way you would talk about a vector in science. One way to do that is to give the angle from the horizontal, θ, and the length of the movement, *r*.

Figure 7.8

Although such a system makes the movement clear, most people don't carry a protractor around to measure the angle of an object's motion. The other system for specifying the motion resolves the translation into its horizontal and vertical components. Every oblique movement is part side-to-side movement and part up-and-down movement. Those are easier to measure, especially when working on the coordinate plane.

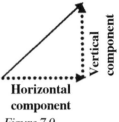

Horizontal component

Figure 7.9

Example 1

The point *P* is translated from its original position at (−7, 3) to a new position at (4, −2). Find the horizontal and vertical components of this translation.

Solution: In order to move from (−7, 3) to (4, −2), point *P* must move 11 units to the right, from an *x*-coordinate of −7 to an *x*-coordinate of 4, and down 5 units, from a *y*-coordinate of 3 to a *y*-coordinate of −2. The horizontal component of the translation is +11, that is, 11 to the right, and the vertical component is −5, or down 5.

When you are working in the coordinate plane, you can designate a translation such as this as $T(x, y) \rightarrow (x + 11, y - 5)$. This notation shows the horizontal and vertical components of the translation, and lets you quickly compute the coordinates of an image point, if you know the coordinates of the preimage (and compute the preimage if you know the image).

Example 2

Find the image of the point (12, −9) under the translation $T(x, y) \rightarrow (x - 5, y + 3)$.

Solution: This translation moves the point 5 units left and 3 units up. The image of (12, −9) will be the point $(12 - 5, -9 + 3)$ or (7, −6).

Example 3

The image of point *P* under the translation $T(x, y) \rightarrow (x - 1, y - 3)$ is *P*′(4, −7). Find the coordinates of the preimage point *P*.

Solution: The translation has moved the point 1 unit left and 3 units down. To move back to the original, or preimage, point, you need to move 1 unit right and 3 units up. Algebraically, you can say $(x - 1, y - 3) = (4, -7)$ so $x - 1 = 4$, meaning $x = 5$, and $y - 3 = -7$ so $y = -4$. The preimage is $P(5, -4)$.

When a translation works on a line, an angle, or a polygon, it slides the figure point by point. Each point moves in exactly the same direction and moves exactly the same distance, so the image is identical to the preimage in size and shape. Translation preserves length and angle measure. The preimage and the image are congruent.

Example 4

The quadrilateral *NOTE* has vertices $N(-3, 1)$, $O(0, 0)$, $T(-1, -4)$, and $E(-5, -1)$. Draw the quadrilateral and its image $N'O'T'E'$ under the translation $T(x, y) \rightarrow (x + 5, y + 2)$. Use the distance formula to verify that $\overline{TE} \cong \overline{T'E'}$.

Solution:

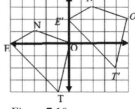

$N' = T(-3, 1) \rightarrow (-3 + 5, 1 + 2) = (2, 3)$

$O' = T(0, 0) \rightarrow (0 + 5, 0 + 2) = (5, 2)$

$T' = T(-1, -4) \rightarrow (-1 + 5, -4 + 2) = (4, -2)$

$E' = T(-5, -1) \rightarrow (-5 + 5, -1 + 2) = (0, 1)$

Figure 7.10

Using the distance formula:

$$TE = \sqrt{\left(-5-(-1)\right)^2 + \left(-1-(-4)\right)^2} = \sqrt{(-4)^2 + (3)^2}$$
$$= \sqrt{16+9} = \sqrt{25} = 5$$

$$T'E' = \sqrt{(0-4)^2 + \left(1-(-2)\right)^2} = \sqrt{(-4)^2 + (3)^2}$$
$$= \sqrt{16+9} = \sqrt{25} = 5$$

Lesson 7-2 Review

Find the image of the point $P(6, -3)$ under each translation.

1. $T(x, y) \rightarrow (x - 4, y + 3)$

2. $T(x, y) \rightarrow (x + 21, y - 17)$

3. $T(x, y) \rightarrow (x, y + 9)$

4. $T(x, y) \rightarrow (x - 9, y)$

5. Given the points $S(0, 3)$, $O(4, 0)$, $N(7, 4)$, and $G(3, 7)$. Graph $\square SONG$ and its image under the translation $T(x, y) \rightarrow (x - 9, y - 10)$, and use the distance formula to verify that $\overline{SO} \cong \overline{S'O'}$.

6. $\triangle D'O'T'$ is the image of $\triangle DOT$ under the translation $T(x, y) \rightarrow (x - 7, y + 5)$. If $D'(-1, -2)$, $O'(1, 5)$, and $T'(5, 2)$, find the preimage of $\triangle D'O'T'$.

Lesson 7-3: Rotations

On a Ferris wheel, the passenger car, or gondola, moves along a circular path. Most Ferris wheel rides have the passenger cars hinged to the frame so that, as the wheel rotates, the car remains upright. When you study rotation in geometry, you're interested in the kind of rotation you'd get if the gondola were held stiffly instead of being hinged.

Translation is the composition of two reflections, over parallel reflecting lines. Rotation is also the composition of two reflections, but

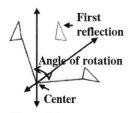

Figure 7.11

over intersecting lines. The point at which the two reflecting lines intersect is the center of the rotation, the center of the circle around which the image moves. The amount of rotation, the distance the image travels, is measured by the angle formed by rays from the center to corresponding points on the preimage and image. As the figure travels around the circle, its size and shape remain the same, so the image of a polygon under a rotation is congruent to the preimage.

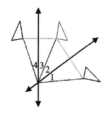

Figure 7.12

The amount of rotation depends on the angle between the two reflecting lines. If you draw lines from the center of rotation to corresponding points of the preimage, first reflection, and final image, four angles are created, which are numbered in Figure 7.12. If you connect preimage and image points, you can prove by congruent triangles that $\angle 1 \cong \angle 2$ and $\angle 3 \cong \angle 4$. The angle between the reflecting lines is equal to m$\angle 2$ + m$\angle 3$.

The angle of rotation is equal to m∠1 + m∠2 + m∠3 + m∠4. The angle of rotation is twice as large as the angle between the reflecting lines.

Example 1

△*RST* is rotated about point *P*, by reflecting first over $\overrightarrow{AB}$ and then over $\overrightarrow{CD}$. Draw the image of △*RST* under this rotation. Find the angle of rotation if m∠*BPC* = 35°.

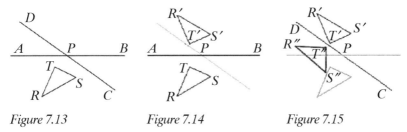

Figure 7.13　　　　　　Figure 7.14　　　　　　Figure 7.15

Solution: Reflect △*RST*, ignoring $\overrightarrow{CD}$ for the moment. Then reflect △*R′S′T′* over $\overrightarrow{CD}$. It may seem that the figure is moving backwards, but don't be disturbed by that. △*R″S″T″* is the image of △*RST* under this rotation. The angle of rotation can be found by drawing ∠*TPT″* and measuring it with a protractor, or by recognizing that the angle of rotation will be twice the size of ∠*BPC*. The angle of rotation is 70°.

Angle of Rotation	Reflection Line 1	Shortcut	Reflection Line 2	Shortcut	Combined Shortcut
90° counter-clockwise	$y = x$	Swap coordinates	y-axis	Change sign of x-coordinate	$(x, y) \rightarrow (-y, x)$
(2, 3) over $y = x$ → (3, 2) over y-axis → (−3, 2)					
180° counter-clockwise	y-axis	Change sign of x-coordinate	x-axis	Change sign of y-coordinate	$(x, y) \rightarrow (-x, -y)$
(1, 4) over y-axis → (−1, 4) over x-axis → (−1, −4)					
270° counter-clockwise (or 90° clockwise)	$y = x$	Swap coordinates	x-axis	Change sign of y-coordinate	$(x, y) \rightarrow (y, -x)$
(3, 5) over $y = x$ → (5, 3) over x-axis → (5, −3)					

Figure 7.16

When a figure is rotated in the coordinate plane, the center of rotation is usually the origin. When that's the case, the reflecting lines may be any lines with *y*-intercepts of zero, but many common angles of rotation can be achieved by combining reflections over the axes and the line $y = x$.

Example 2

Quadrilateral *ABCD* has vertices *A*(3, 0), *B*(4, 2), *C*(2, 5), and *D*(1, 1). Draw the image of *ABCD* under a rotation about the origin of:

a. 90°   b. 180°   c. 270°

Solution:

a. The shortcut to rotate a point 90° is $(x, y) \rightarrow (-y, x)$.
   $A(3, 0) \rightarrow A'(0, 3), B(4, 2) \rightarrow B'(-2, 4), C(2, 5) \rightarrow C'(-5, 2)$, and $D(1, 1) \rightarrow D'(-1, 1)$

b. The shortcut to rotate a point 180° is $(x, y) \rightarrow (-x, -y)$.
   $A(3, 0) \rightarrow A'(-3, 0), B(4, 2) \rightarrow B'(-4, -2), C(2, 5) \rightarrow C'(-2, -5)$, and $D(1, 1) \rightarrow D'(-1, -1)$

c. The shortcut to rotate a point 270° is $(x, y) \rightarrow (y, -x)$.
   $A(3, 0) \rightarrow A'(0, -3), B(4, 2) \rightarrow B'(2, -4), C(2, 5) \rightarrow C'(5, -2)$, and $D(1, 1) \rightarrow D'(1, -1)$

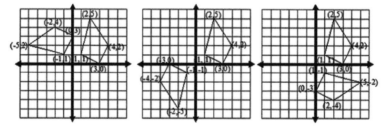

Figure 7.17

Lesson 7-3 Review

1. Find the image of □*QRST* rotated 110° about point *P*.

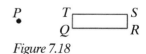

Figure 7.18

2. △*ABC* in Figure 7.19 is rotated about point *P*, by reflecting first over $\overleftrightarrow{XY}$ and then over $\overleftrightarrow{ST}$. Draw the image of △*ABC* under this rotation.

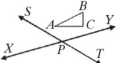

Figure 7.19

3. Each polygon is rotated by reflecting first over l_1 and then over l_2. Find the angle of rotation.

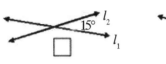 

Figure 7.20

4. Draw the image of each polygon under a rotation about the origin with the angle of rotation shown.

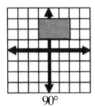 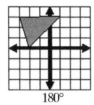

90°   180°

Figure 7.21

Lesson 7-4: Dilations

The transformation known as a dilation differs from the others because it does not preserve the size of the preimage. Instead, it enlarges or reduces the size proportionally, so that the image and the preimage are similar figures.

To understand dilations, it may help to think about a projector throwing an image onto a screen. The job of the projector is to take a small picture and produce a larger copy of it on the screen. Think of the small picture as the preimage and the larger copy as the image. To accomplish its job the projector must have a light source from which rays of light stretch out to the screen. The light source is the center of the dilation. The greater the distance between the projector and the screen, the longer the rays of light, and the larger the image will appear. In a dilation, the scale factor tells you about the length of the rays, and therefore about the size of the image.

Figure 7.22

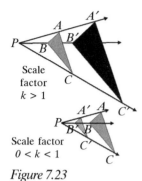

Scale factor
k > 1

Scale factor
0 < k < 1

Figure 7.23

If △*ABC* is the preimage, and the dilation is centered at point *P*, imagine rays of light from point *P* moving out to each of the vertices of the triangle. These give you rays $\overrightarrow{PA}$, $\overrightarrow{PB}$, and $\overrightarrow{PC}$. If the image triangle (△*A′B′C′*) is to be larger than △*ABC*, it will be farther away from point *P*, and *PA′* will be greater than *PA*, *PB′* greater than *PB*, and *PC′* greater than *PC*. If, on the other hand, the image is to be smaller than the preimage, it will be closer to *P*, and so *PA′*, *PB′*, and *PC′* will be smaller than *PA*, *PB*, and *PC*, respectively. The scale factor acts as a multiplier, telling you how much larger or smaller. If the scale factor is 2, *PA′*, *PB′*, and *PC′* are twice as long as *PA*, *PB*, and *PC*. If the scale factor is $\frac{1}{3}$, *PA′*, *PB′*, and *PC′* are one third as long as *PA*, *PB*, and *PC*.

Example 1

Find the image of △*XYZ* under a dilation centered at point *M* if the scale factor is $k = \frac{1}{2}$.

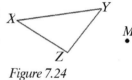

Figure 7.24

Solution: Draw $\overrightarrow{MX}$, $\overrightarrow{MY}$, and $\overrightarrow{MZ}$, and measure the lengths *MX*, *MY*, and *MZ*. Then $MX' = \frac{1}{2}MX$, $MY' = \frac{1}{2}MY$, and $MZ' = \frac{1}{2}MZ$. Mark *X′* on $\overrightarrow{MX}$, *Y′* on $\overrightarrow{MY}$, and *Z′* on $\overrightarrow{MZ}$ and connect to form the image triangle △*X′Y′Z′*

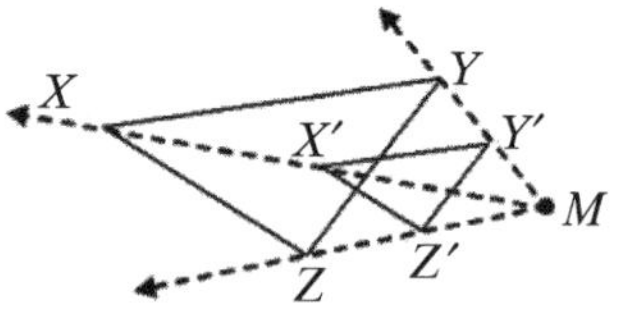

Figure 7.25

Example 2

Find the image of □*MNOP* under a dilation centered at point *P* if the scale factor is *k* = 1.5.

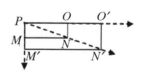

Figure 7.26

Solution: Because the center of the dilation is a vertex of □*MNOP*, the ray $\overrightarrow{PP}$ is a single point, so *P* and *P′* are the same point. $\overrightarrow{PO}$ and $\overrightarrow{PM}$ run along the sides of the rectangle, and $\overrightarrow{PN}$ is a diagonal. Measure *PO, PM,* and *PN,* and multiply each by 1.5 to find the length of *P'O', P'M',* and *P'N'.* Connect *P', O', N',* and *M'* to form the image $\square M'N'O'P'$.

In the coordinate plane, the center of the dilation is usually the origin. If (x, y) is a point in the preimage, and k is the scale factor for a dilation centered at the origin, then the corresponding image point is (kx, ky).

Example 3

Find the image of pentagon *ABCDE* under a

dilation with scale factor $k = \dfrac{1}{3}$ centered at the

origin.

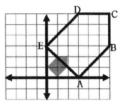

Figure 7.27

Solution: Identify the coordinates of each vertex.

$A(3, 0)$, $B(6, 3)$, $C(6, 6)$, $D(3, 6)$ and $E(0, 3)$

The vertices of the image are:

$$A': \left(\frac{1}{3} \cdot 3, \frac{1}{3} \cdot 0\right) = (1, 0) \qquad C': \left(\frac{1}{3} \cdot 6, \frac{1}{3} \cdot 6\right) = (2, 2)$$

$$B': \left(\frac{1}{3} \cdot 6, \frac{1}{3} \cdot 3\right) = (2, 1) \qquad D': \left(\frac{1}{3} \cdot 3, \frac{1}{3} \cdot 6\right) = (1, 2)$$

$$E': \left(\frac{1}{3} \cdot 0, \frac{1}{3} \cdot 3\right) = (0, 1)$$

Plot the image points and connect to form pentagon *A'B'C'D'E'*.

The image of a polygon under dilation is similar to the preimage polygon, and the ratio of similitude is equal to the scale factor.

Example 4

Find the center and scale factor of the dilation if $\triangle R'S'T'$ is the image of $\triangle RST$, $RT = 24$, and $R'T' = 16$.

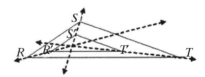

Figure 7.28

Solution: To find the center of the

dilation draw $\overrightarrow{RR'}$, $\overrightarrow{SS'}$, and $\overrightarrow{TT'}$

and extend until the three lines intersect. The point of intersection is the center of the dilation. This dilation will have a scale factor less than 1 because the image is smaller than the preimage. The ratio of the known side of the image to the known side of the

preimage is $\frac{16}{24} = \frac{2}{3}$, so the ratio of similitude of the triangles and

the scale factor of the dilation is $\frac{2}{3}$.

Lesson 7-4 Review

1. In Figure 7.29, find the image of $\triangle XYZ$ under a

   dilation centered at point M if the scale factor is $k = \frac{3}{4}$.

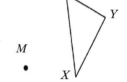

Figure 7.29

2. In Figure 7.30, find the image of $\triangle XYZ$ under a dilation centered at point Y if the scale factor $k = 2.5$.

3. $\triangle ABC$ has vertices $A(-2, -1)$, $B(1, 3)$ and $C(3, -2)$. Find the image of $\triangle ABC$ under a dilation centered at the origin with a scale factor of $k = 3$.

4. $\triangle R'S'T'$ is the image of $\triangle RST$ under a dilation

   with scale factor $k = \frac{4}{3}$. Find the length of $\overline{ST}$ if $ST' = 9$.

Figure 7.30

5. Find the center of the dilation represented in Figure 7.31.

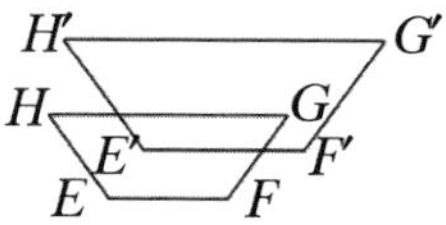

Figure 7.31

Lesson 7-5: Symmetry

Whether in nature or in design, an object that is symmetric has a regularity, a pattern to its design or arrangement, that we can describe by reflection or rotation. The wings of a butterfly (mirror images of one another) have reflection, or line, symmetry. The crystalline patterns of snowflakes are examples of rotational or point symmetry.

Figure 7.32

A figure has reflection symmetry if you can find a line to act as a reflecting line, and the image of the figure under that reflection is identical to the preimage. Isosceles triangle △*ABC* is symmetric about the line $\overleftrightarrow{BD}$ that bisects its vertex angle and is the perpendicular bisector of the base $\overline{AC}$ at point *D*. If △*ABC* is reflected over $\overleftrightarrow{BD}$, △*C′B′A′* is identical to △*ABC*.

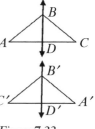

If a figure has reflection symmetry, the reflecting line is called a **line of symmetry**. A polygon may have one line of symmetry, many lines of symmetry, or none at all.

Figure 7.33

Example 1

Determine whether each figure has reflection symmetry, and, if so, find any lines of symmetry.

Figure 7.34

Solution:

  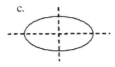

Figure 7.35

a. No reflection symmetry. (Because its diagonals will divide it into congruent triangles, many people assume they are lines of symmetry, but when reflected over the diagonal, as shown, the figure is not identical to the original.)

b. The heart has one line of symmetry.

c. The ellipse has two lines of symmetry.

A figure has rotation symmetry if you can determine some rotation, both center and angle, for which the image is identical to the preimage. Just as a polygon that has reflection symmetry may have multiple lines of symmetry, a polygon may have rotational symmetry for more than one angle. A regular hexagon has rotational symmetry because each rotation of 60°, or a multiple of 60°, about the center, produces an image congruent to the original.

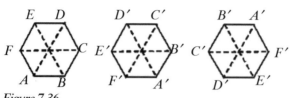

Figure 7.36

Example 2

Determine whether each of the following figures has rotational symmetry. If so, show the center and angle of rotation that produces the identical image.

a.   b.   c.

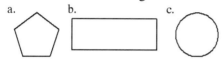

Figure 7.37

Solution:

a. The regular pentagon has rotational symmetry about point P. The first identical image is produced by making one-fifth of a full rotation, or a rotation of

  $\dfrac{360}{5} = 72°$. Images produced by rotations of 144°, 216°, and 288° are also identical to the original.

b. The rectangle can be rotated 180° about point P to produce an image identical to the original.

c. The circle has rotational symmetry about its center for any angle of rotation. Every diameter is a line of symmetry.

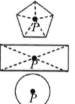

Figure 7.38

A polygon may have both reflection and rotation symmetry. The regular hexagon, for example, has three lines of symmetry, as shown in Figure 7.36, as well as rotational symmetry.

Example 3

Determine what type of symmetry, if any, is present in each figure.

a.   b.   c.   d.

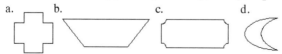

Figure 7.39

Solution:

a. The figure has reflection symmetry with four lines of symmetry, as shown. It also has rotation symmetry, about point P, with rotation angles of 90°, 180°, or 270°.

Figure 7.40

b. There is one line of symmetry, so the trapezoid has reflection symmetry, but not rotational symmetry.

Figure 7.41

c. The figure has reflection symmetry, with two lines of symmetry, as shown. It has rotation symmetry about point *P* with a rotation angle of 180°.

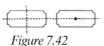

Figure 7.42

d. The figure has reflection symmetry but not rotational symmetry. The line of reflection is shown.

Figure 7.43

Lesson 7-5 Review

Determine what type of symmetry, if any, is present in each polygon. If the figure has reflection symmetry, specify the reflecting line. If the figure has rotational symmetry, give the center and angle of rotation.

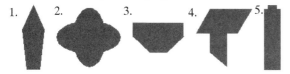

Figure 7.44

Answer Key
Lesson 7-1

1.

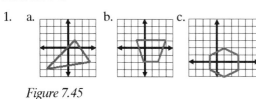

Figure 7.45

Lesson 7-2

1. (2,0)

2. (27,−20)

3. $(6, 6)$

4. $(-3, -3)$

5. $S'(-9, -7), O'(-5, -10), N'(-2, -6)$, and $G'(-6, -3)$.

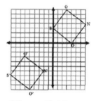

$$SO = \sqrt{(0-4)^2 + (3-0)^2} = \sqrt{25} = 5$$

$$S'O' = \sqrt{(-9-(-5))^2 + (-7-(-10))^2}$$
$$= \sqrt{25} = 5$$

Figure 7.46

6. $D: x - 7 = -1 \Rightarrow x = 6$ and $y + 5 = -2 \Rightarrow y = -7$ so $D(6, -7)$.

$O: x - 7 = 1 \Rightarrow x = 8$ and $y + 5 = 5 \Rightarrow y = 0$ so $O(8, 0)$.

$T: x - 7 = 5 \Rightarrow x = 12$ and $y + 5 = 2 \Rightarrow y = -3$ so $T(12, -3)$.

Lesson 7-3

1.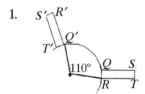

Figure 7.47

2.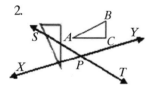

Figure 7.48

3. $2(15°) = 30°$ angle of rotation

   $2(70°) = 140°$ angle of rotation

4.

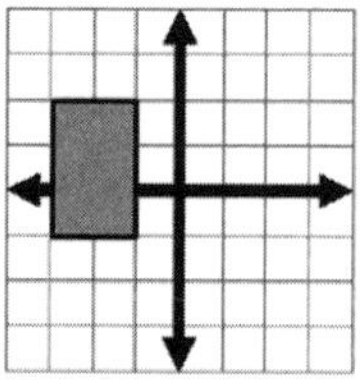

Figure 7.49 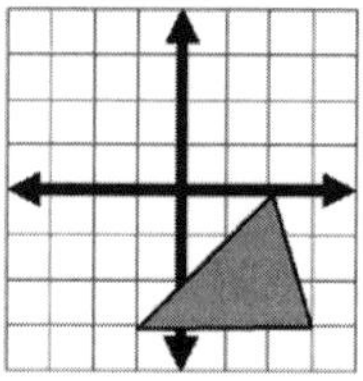   *Figure 7.50*

Lesson 7-4

1.

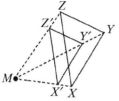

Figure 7.51

2.

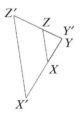

Figure 7.52   *Figure 7.53*

3. $A'(-6,-3), B'(3,9)$ and $C'(9,-6)$

4. $\dfrac{ST}{S'T'} = \dfrac{4}{3} \Rightarrow \dfrac{x}{9} = \dfrac{4}{3} \Rightarrow 3x = 36 \Rightarrow x = 12$.

$\overline{ST}$ has a length of 12.

5.   Point P is the center of the dilation shown.

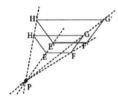

Figure 7.54

Lesson 7-5

1..

Reflection symmetry, one line of symmetry

2. 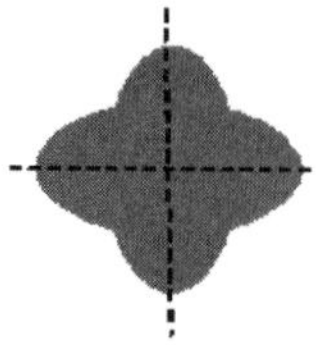   Reflection symmetry, two lines of symmetry, and rotation symmetry about P, 90°, 180°, or 270°.

3. 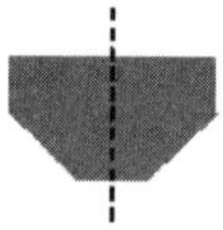   Reflection symmetry, one line of symmetry

4.    No symmetry

5.    Reflection symmetry, one line of symmetry

Figure 7.55

Right Triangles

Lesson 8-1: Working With Right Triangles

A right triangle is a triangle that contains one right angle. Because the three angles of any triangle add to 180°, the two angles other than the right angle will be complementary, acute angles.

The **Pythagorean Theorem** says that if a and b are the lengths of the legs of a right triangle and c is the length of the hypotenuse, then $a^2 + b^2 = c^2$. The Pythagorean relationship can be used to find the length of a side of a right triangle when the other two are known, or to verify that a triangle actually is a right triangle.

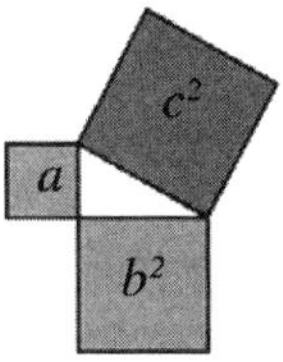

Figure 8.1

One proof of this theorem uses four identical copies of the right triangle. All four triangles have the same area. Position them as shown and you have a large square with each side of length $a + b$, and inside it, tilted, a square with sides of length c.

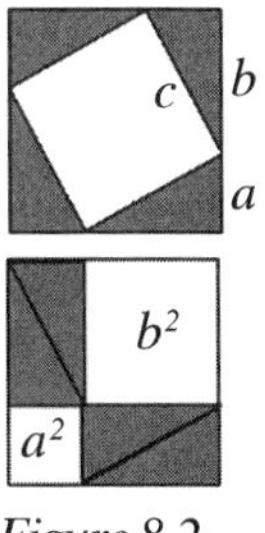

You can show, visually, that the white area, c^2, is equal to $a^2 + b^2$, by holding the frame of the large square and rearranging the triangles inside it. Slide pairs of triangles together

Figure 8.2

and the remaining white area is two squares, one formed by the shorter sides of the triangles, with area a^2, and one formed by the larger legs, with area b^2.

Just so that doesn't seem to be sleight of hand, you can prove it algebraically. The area of the large square is $(a + b)^2$, because each side is $a + b$, and $(a + b)^2 = a^2 + 2ab + b^2$. The same region is equal to the area of the tilted white square plus the areas of the four triangles. The white square had an area of c^2, and each of the four triangles has an area of

$\frac{1}{2}ab$, because the legs can be used as the base and the height. Finding the area that way, you get $c^2 + 4\left(\frac{1}{2}ab\right) = c^2 + 2ab$. The area of the square is the same, either way, so $a^2 + 2ab + b^2 = c^2 + 2ab$. Subtract $2ab$ from both sides and $a^2 + b^2 = c^2$.

Example 1

If the base of a 28-foot ladder is placed 5 feet from the base of a wall, how far up on the wall will the ladder reach?

Solution: Assume that the wall and the floor form a right angle, and the ladder forms the hypotenuse of the right triangle. Then $a = 5$, b is unknown, and $c = 28$.

$a^2 + b^2 = c^2$

$5^2 + b^2 = 28^2$

$25 + b^2 = 784$

$b^2 = 759$

$b \approx 27.5$

The ladder will reach approximately 27.5 feet up the wall.

Example 2

Martin's home has a deck that appears to be a rectangle. Martin measures two sides and the diagonal, and finds they are 6 feet, 8 feet, and 10 feet, 3 inches. Is Martin's deck actually a rectangle?

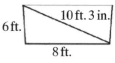

Figure 8.3

Solution: If the deck is truly a rectangle, the diagonal will divide it into two right triangles, and the measurements of those triangles should fit the Pythagorean theorem. The corner to corner measurement of 10 feet, 3 inches is 10.25 feet, so check to see if $6^2 + 8^2 = (10.25)^2$. On the left side, $6^2 + 8^2 = 36 + 64 = 100$, but on the right, $(10.25)^2 = 105.0625$. Because $100 \neq 105.0625$, the triangle is not a right triangle and the deck is not a rectangle.

The moment you identify two triangles as right triangles, you know that one pair of corresponding angles—the right angles—are congruent.

All that's left to show is a pair of acute angles congruent and the requirements of AA are satisfied. The result is that all right triangles that contain an acute angle of 30° are similar to one another, and all right triangles
that contain an acute angle of 18° are similar to one another. Right triangles can be organized into families, according to their acute angles: the
1° family, the 2° family, and so on.

Imagine you take a family of similar right triangles—say the 43°
family—and stack them on top of one another so that the acute angles
that give the family its name are aligned. The hypotenuse
and one of the legs form the 43° angle. In Figure 8.4, you
can see that the hypotenuse of each triangle in the family falls on the same line, and that the legs adjacent to
the 43° angle stack on top of one another. The legs that
fall opposite the 43° angles form parallels. Because all

Figure 8.4

the triangles in the family are similar, if you pick out two of them, you
can be sure that the corresponding sides are in proportion.

$$\frac{\text{hypotenuse \#1}}{\text{hypotenuse \#2}} = \frac{\text{adjacent \#1}}{\text{adjacent \#2}} = \frac{\text{opposite \#1}}{\text{opposite \#2}}$$

Example 3

$\triangle OCN$ and $\triangle OEA$ are similar right triangles.
If $OC = 6$, $OE = x$, $ON = 2x$, and $OA = 27$, find NA.

Figure 8.5

Solution: $\triangle OCN \sim \triangle OEA$ so corresponding sides are
in proportion. Using the sides in the given information,
you can write $\dfrac{OC}{OE} = \dfrac{ON}{OA}$ and substitute to get $\dfrac{6}{x} = \dfrac{2x}{27}$.
Cross-multiplying gives you $2x^2 = 162$, so $x^2 = 81$ and $x = 9$. $OE = 9$
and $ON = 18$, so $NA = OA - ON = 27 - 18 = 9$.

Example 4

When water is poured into a cone-shaped drinking cup, the surface
of the water forms a circle, and the radius of the circle changes as
the depth of the water changes. When the cone is full, the water is
6 inches deep and the radius of the circle is 1 inch. What is the
radius of the circle when the water is 4 inches deep?

Solution: If you draw lines for the depth of the water and lines for the radii of the circles, you'll see they form right triangles with the slanted side of the cone as the hypotenuse. Use similar right triangles to find the missing radius. $\dfrac{\text{depth}}{\text{depth}} = \dfrac{\text{radius}}{\text{radius}}$ so $\dfrac{6}{4} = \dfrac{1}{x}$.

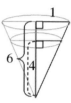

Cross-multiplying gives you $6x = 4$, which means $x = \dfrac{2}{3}$.

The radius of the circle is two-thirds of an inch when the water is 4 inches deep.

Figure 8.6

Lesson 8-1 Review

1. Find the measure of the unmarked angle in each right triangle.

a.
b.
c.

Figure 8.7

2. Find the missing side in each triangle.

a.
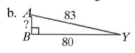
b.

c.

Figure 8.8

3. $\triangle STR$ and $\triangle EAM$ are right triangles, with $\overline{ST} \perp \overline{TR}$ and $\overline{EA} \perp \overline{AM}$. If $\angle R \cong \angle M$, $ST = 5$, $SR = 13$, $EA = 15$ and $AM = 36$. Find TR and EM.

4. At a certain time of day, Marisa notices that the tip of the shadow cast by her house and the tip of the shadow cast by the crab apple tree in her yard land in exactly the same place. After making a sketch, she realizes that there are similar right triangles at work. The tips of the shadows are 12 feet from the base of the tree. The tree, which is planted 18 feet from the center of the house, is 10 feet tall. How tall is Marisa's house?

Figure 8.9

Lesson 8-2: More Splitting

You've learned several theorems about ways to split the sides of triangles proportionally. Those apply to all triangles: right, acute, or obtuse. There is another splitter theorem that applies only to right triangles, in which an altitude from the right angle divides the hypotenuse.

The altitude to the hypotenuse creates two triangles, which are both similar to the original triangle, and which are similar to one another. All of the proportions you'll need come from these similar triangles.

If $\triangle PDN$ is a right triangle with right angle $\angle D$, and altitude $\overline{DO}$ is drawn to hypotenuse $\overline{PN}$, the right angle in the small triangle, $\angle POD$, is congruent to the right angle in the big triangle, $\angle PDN$. $\angle P$ is in both triangles, so that will correspond to itself, and that leaves $\angle PDO \cong \angle N$. Find the alignment between the other small triangle and the original ($\angle DON \cong \angle PDN$, $\angle N \cong \angle N$, and $\angle NDO \cong \angle P$) and you can transfer the correspondence: $\triangle PDN \sim \triangle POD \sim \triangle DON$. Figure 8.11 summarizes the corresponding parts.

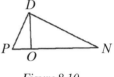

Figure 8.10

Triangle	$\triangle PDN$	$\triangle POD$	$\triangle DON$
Angles	$\angle P$	$\angle P$	$\angle NDO$
	$\angle N$	$\angle PDO$	$\angle N$
Right angle	$\angle PDN$	$\angle POD$	$\angle DON$
Legs	$\overline{DN}$	$\overline{DO}$	$\overline{ON}$
	$\overline{PD}$	$\overline{PO}$	$\overline{DO}$
Hypotenuse	$\overline{PN}$	$\overline{PD}$	$\overline{DN}$

Figure 8.11

Example 1

Write the similarity statements for the three triangles in the figure.

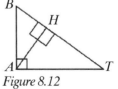

Figure 8.12

Solution: Start with $\triangle BAT$ and $\triangle BHA$. Right angles are congruent so $\angle BAT \cong \angle BHA$, and $\angle B \cong \angle B$, so $\triangle BAT \sim \triangle BHA$. Going to the other small triangle, $\angle BAT \cong \angle THA$ and $\angle T \cong \angle T$, so $\triangle BAT \sim \triangle HAT$. By the transitive property, $\triangle BHA \sim \triangle HAT$.

One theorem says that the altitude is the mean proportional between the two pieces of the hypotenuse. Because $\triangle POD \sim \triangle DON$, $\frac{PO}{DO} = \frac{DO}{ON}$.

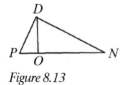

Figure 8.13

An altitude to the hypotenuse of a right triangle splits the hypotenuse into two pieces so that:

$$\frac{\text{piece \#1}}{\text{altitude}} = \frac{\text{altitude}}{\text{piece \#2}}.$$

The altitude $\overline{DO}$ is the mean proportional, and $\overline{PO}$ and $\overline{ON}$ are the segments of the hypotenuse. The altitude to the hypotenuse is the mean proportional between the segments of the hypotenuse.

Example 2

In right triangle $\triangle SIN$, with right angle at I, altitude $\overline{IK}$ is drawn to the hypotenuse. If $SK = 9$ and $KN = 16$, find IK.

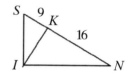

Figure 8.14

Solution: IK is the mean proportional between SK and KN, so $\dfrac{SK}{IK} = \dfrac{IK}{KN}$. Substitute to find $\dfrac{9}{x} = \dfrac{x}{16}$. Cross-multiplying gives you $x^2 = 144$, so $x = 12$. The length of the altitude is 12 units.

Either leg of the original right triangle is the mean proportional between the whole hypotenuse and the segment of the hypotenuse nearest the leg. Focus on $\triangle POD$ and $\triangle PDN$, and write the proportion:

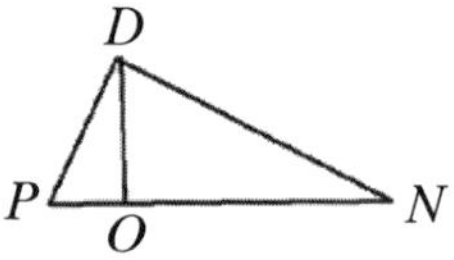

D

P O N

Figure 8.15

$$\frac{\text{leg}}{\text{leg}} = \frac{\text{hypotenuse}}{\text{hypotenuse}} \quad \text{or} \quad \frac{PO}{PD} = \frac{PD}{PN}.$$

PD, the length of one leg of the original triangle, is the mean proportional between PO and PN. $\overline{PO}$ is the segment of the hypotenuse nearest to $\overline{PD}$, and $\overline{PN}$ is the whole hypotenuse. If you look at $\triangle DON$ and $\triangle PDN$ instead, the proportion is $\frac{ON}{DN} = \frac{DN}{PN}$. It puts DN, the length of the other leg of the right triangle, as the mean proportional between ON and PN. ON is the length of the segment of the hypotenuse nearest to DN. You can choose either leg of the original right triangle, and it will be the mean proportional between the hypotenuse and the part of the hypotenuse closest to that leg.

Example 3

In right triangle $\triangle PAL$, altitude $\overline{AI}$ is drawn to hypotenuse $\overline{PL}$. If $PI = 14x - 2, AP = 6$, and $IL = 16x$, find the length of the altitude and the length of the hypotenuse.

Figure 8.16

Solution: AP is the mean proportional between PI and PL, but PL is not given. It can be expressed as $PI + IL$, however, so $PL = 14x - 2 + 16x = 30x - 2$.

$$\frac{PI}{AP} = \frac{AP}{PL}$$

$$\frac{14x-2}{6} = \frac{6}{30x-2}$$

$$(14x-2)(30x-2) = 36$$

$$420x^2 - 60x - 28x + 4 = 36$$

$$420x^2 - 88x - 32 = 0$$

$$105x^2 - 22x - 8 = 0$$

$$(21x+4)(5x-2) = 0$$

Setting $21x + 4$ equal to 0 yields a negative value of x, which makes no sense in this situation, but if $5x - 2 = 0$, then $x = 0.4$, and the hypotenuse has a length of $PL = 30(0.4) - 2 = 12 - 2 = 10$. The hypotenuse is 10 units long. To find the length of the altitude, remember that the altitude is the mean proportional between the segments of the hypotenuse. Because x is known to be 0.4, $PI = 14(0.4) - 2 = 3.6$, and $IL = 16(0.4) = 6.4$.

$$\frac{PI}{AI} = \frac{AI}{IL}$$

$$\frac{3.6}{y} = \frac{y}{6.4}$$

$$y^2 = 23.04$$

$$y = 4.8$$

The length of the altitude is 4.8 units.

Lesson 8-2 Review

1. In right triangle ΔRST, altitude $\overline{SU}$ is drawn from the right angle at S to the hypotenuse $\overline{RT}$. If $RU = 9$ and $SU = 15$, find UT.

2. In a certain right triangle (Figure 8.17), the altitude to the hypotenuse divides the hypotenuse into two congruent segments. Calculate a and b to show that the right triangle is isosceles.

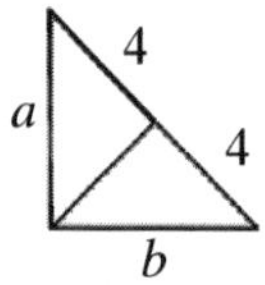

Figure 8.17

3. In right triangle ΔABC, altitude $\overline{BD}$ is drawn to hypotenuse $\overline{AC}$. If $AD = 5$, $BD = 7$, and $BC = 9$, find AB and DC.

Lesson 8-3: Trigonometry

The word *trigonometry* means triangle measurement, and the study of trigonometry depends on families of similar right triangles. Pick any two triangles in a family and you know corresponding sides are in proportion, so you can write a proportion in which:

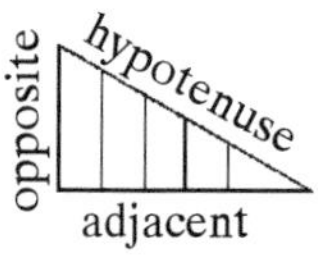

Figure 8.18

$$\frac{\text{opposite}}{\text{opposite}} = \frac{\text{adjacent}}{\text{adjacent}} \text{ or } \frac{\text{adjacent}}{\text{adjacent}} = \frac{\text{hypotenuse}}{\text{hypotenuse}} \text{ or } \frac{\text{opposite}}{\text{opposite}} = \frac{\text{hypotenuse}}{\text{hypotenuse}}$$

Exchange the means, and these proportions become

$$\frac{\text{opposite}}{\text{adjacent}} = \frac{\text{opposite}}{\text{adjacent}}, \frac{\text{adjacent}}{\text{hypotenuse}} = \frac{\text{adjacent}}{\text{hypotenuse}}, \text{ and } \frac{\text{opposite}}{\text{hypotenuse}} = \frac{\text{opposite}}{\text{hypotenuse}}$$

In ΔABC

$$\frac{\text{side opposite } \angle A}{\text{side adjacent to } \angle A} = \frac{6}{8} = \frac{3}{4}$$

$$\frac{\text{side opposite } \angle A}{\text{hypotenuse}} = \frac{6}{10} = \frac{3}{5}$$

$$\frac{\text{side adjacent to } \angle A}{\text{hypotenuse}} = \frac{8}{10} = \frac{4}{5}$$

In ΔXYZ

$$\frac{\text{side opposite } \angle X}{\text{side adjacent to } \angle X} = \frac{9}{12} = \frac{3}{4}$$

$$\frac{\text{side opposite } \angle X}{\text{hypotenuse}} = \frac{9}{15} = \frac{3}{5}$$

$$\frac{\text{side adjacent to } \angle X}{\text{hypotenuse}} = \frac{12}{15} = \frac{4}{5}$$

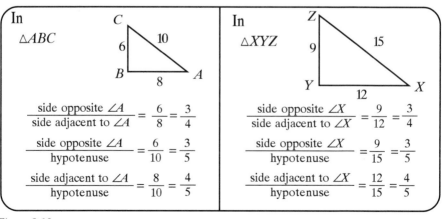

Figure 8.19

If △*ABC* is a right triangle with sides of 6, 8, and 10, and △*XYZ* is similar to △*ABC*, with ∠*A* ≅ ∠*X* and a scale factor of $\frac{3}{2}$, the sides of △*XYZ* will be 9, 12, and 15.

Once you realize that these ratios are the same for all the triangles in a particular family, it makes sense to give each of the ratios a name. Originally, people made lists or tables of the ratios for different families. If you knew your triangle was a part of the 23° family, you could go to a chart and look up the value of the ratio you wanted to use. You might see in the table that the ratio of the opposite side to the adjacent side for an angle of 23° was approximately .4245. If your triangle had an adjacent side of length 10, you knew that $\frac{opposite}{10} = .4245$, so your opposite side was going to be 4.245. Although you can still find tables in some textbooks (or in reference books) to look up values of the different ratios, today's scientific calculators are equipped with keys that will return the value of the ratio when you input the measure of the angle.

Figure 8.20

The ratio of the side opposite an acute angle (call it ∠*A*) to the hypotenuse of the triangle is called the **sine of ∠A.** *Sine* is commonly abbreviated as *sin*, so you'll see sin∠*A* or sin *A* written to mean $\frac{\text{side opposite } \angle A}{\text{hypotenuse}}$.

If you know the length of the opposite side and you want to know the hypotenuse, or if you know the hypotenuse and want to know the length of the opposite side, the sine ratio for that angle family will let you set up an equation.

Example 1

In right triangle △*EAR*, ∠*R* measures 37° and hypotenuse $\overline{ER}$ is 4 cm long. Find the length of $\overline{EA}$.

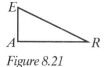

Figure 8.21

Solution: You know the hypotenuse and want to find the length of the side opposite ∠*R*, so

$\frac{\text{opposite}}{\text{hypotenuse}} = \sin \angle R$. Putting in the values you know, the equation

becomes $\frac{x}{4} = \sin 37°$. At this point, you could turn to a table that lists the values of the trigonometric ratios, and find that sin 37° is approximately 0.6018. If your calculator is equipped with keys for the trigonometric ratios, you could use the calculator to find the value,

Angle	Sin	Cos	Tan
36°	.5878	.8090	.7265
37°	.6018	.7986	.7336
38°	.6157	.7880	.7813

Figure 8.22

instead of a table. Then $\dfrac{x}{4} = 0.6018$,

and multiplying both sides by 4 tells you that $x = 2.4072$. The side opposite the 37° angle measures about 2.4 cm.

Example 2

A forest ranger spots a fire 2 miles directly north of his observation post. Another observation post, directly east of the first, also sees the fire. The angle formed by the east-west line connecting the observation towers and the eastern tower's line of sight to the fire is $\angle WEF$ and measures 40°. How far is the fire from the eastern post?

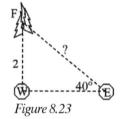

Figure 8.23

Solution: You're looking for the hypotenuse of

$$\triangle WEF. \quad \sin 40° = \frac{\text{opposite}}{\text{hypotenuse}} = \frac{2}{x}.$$

From a calculator or a table, you can find that $\sin 40° \approx 0.6428$, so

$0.6428 = \dfrac{2}{x}$. Cross-multiplying, you get $0.6428x = 2$, so

$x = \dfrac{2}{0.6428} \approx 3.1114$. The fire is approximately 3.11 miles from the eastern tower.

Sometimes you know what the ratio of the two sides is, but you want to know the angle family to which your triangle belongs. If you know that the side opposite $\angle A$ measures 8 inches and the hypotenuse of your triangle measures 16 inches, you know that $\sin \angle A = \frac{8}{16} = \frac{1}{2}$, but what is the measure of $\angle A$? What angle has a sine ratio equal to one-half? What you're looking for is called the **arcsin** or **inverse sine** of $\frac{1}{2}$. When you see

arcsin $\frac{1}{2}$ or $\sin^{-1}\left(\frac{1}{2}\right)$, you should think *the angle whose sine is* $\frac{1}{2}$. If you are using a calculator to find trigonometric ratios, you should see the inverse sine symbol written above the SIN key. To find the angle whose sine is a certain number, use your 2nd or INV key followed by the SIN key. If you are using a table, you will need to look carefully down the sin column until you find $\frac{1}{2}$, or the value in the col-

umn closest to $\frac{1}{2}$, then go across to see what angle has that sine.

Figure 8.24

Example 3

In right triangle $\triangle TOE$, leg $\overline{TO}$ measures 9 meters and hypotenuse $\overline{TE}$ measures 12 meters. Find the measure of $\angle E$.

Solution: $\overline{TO}$ is the leg opposite $\angle E$ and $\overline{TE}$ is the hypotenuse, so $\dfrac{TO}{TE} = \sin\angle E$. The ratio is

$\dfrac{9}{12} = \dfrac{3}{4} = 0.75$, so you want to know what angle has a sine of 0.75. If you look through the sin column of a trig table, you can see that 0.75 would fall between 0.7431 and 0.7547, so the angle whose sine is 0.75 is an angle between 48° and 49°, closer to 49°, because 0.75 is closer to 0.7547 than to 0.7431. If you use a calculator, you find the angle is 48.59°.

Angle	Sin
45°	.7071
46°	.7193
47°	.7314
48°	.7431
49°	.7547
50°	.7660

$\sin^{-1}(.75)$
48.59037789

Figure 8.25

The name cosine is given to the ratio of the adjacent side to the hypotenuse. The **cosine of $\angle P$** is abbreviated cos $\angle P$, and in Figure 8.26, $\cos\angle P = \frac{HP}{IP}$.

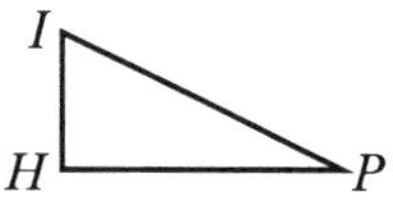

The word *cosine* is a compression of *complementary sine*. It refers to the fact that the cosine of an acute angle is the sine of its complement. In $\triangle HIP$, $\sin\angle P = \frac{HI}{IP}$ and

Figure 8.26

$\cos\angle P = \frac{HP}{IP}$. But there is another acute angle in $\triangle HIP$ (namely $\angle I$) and

$\angle I$ and $\angle P$ are complementary. It turns out that $\sin \angle I = \frac{HP}{IP}$ and $\cos \angle I = \frac{HI}{IP}$, so $\sin \angle I = \cos \angle P$, and $\cos \angle I = \sin \angle P$. The cosine of one angle is the sine of its complement.

Example 4

In right triangle $\triangle LEG$, $\angle E$ is a right angle. $LE = 3$ cm, $EG = 4$ cm, and $LG = 5$ cm. Find:

a. $\sin \angle L$ b. $\cos \angle L$ c. $\sin \angle G$ d. $\cos \angle G$

Solution: The hypotenuse is LG, which is 5 cm long. For $\angle L$, the opposite is EG and the adjacent is LE.

a. $\sin \angle L = \dfrac{4}{5}$ b. $\cos \angle L = \dfrac{3}{5}$

For $\angle G$, the opposite is LE and the adjacent is EG.

c. $\sin \angle G = \dfrac{3}{5}$ d. $\cos \angle G = \dfrac{4}{5}$

Example 5

In rectangle $\square NOSE$, diagonal $\overline{NS}$ makes an angle of 22° with side $\overline{NO}$. If $NO = 12$ cm, find the length of diagonal $\overline{NS}$ and the length of side $\overline{SO}$.

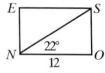

Figure 8.27

Solution: The diagonal divides the rectangle into two right triangles, so you can use right triangle $\triangle NOS$, with $\angle SNO$ as your acute angle.

$\cos \angle NSO = \dfrac{NO}{NS}$ so $\cos 22° = \dfrac{12}{x}$. From a calculator or a table, you can find that $\cos 22° = 0.9272$. The equation becomes $0.9272 = \dfrac{12}{x}$ and cross-multiplying gives you $0.9272x = 12$. Dividing both sides by 0.9272 tells you that $NS \approx 12.94$ cm. To find SO, use the equation $\sin 22° = \dfrac{SO}{NS} = \dfrac{y}{12.94}$. Because $\sin 22° = 0.3746$, $0.3746 = \dfrac{y}{12.94}$, and $y = (0.3746)(12.94)$. Side SO is approximately 4.85 cm.

Example 6

The safety sticker on Shemar's ladder recommends placing it so that the foot of the ladder makes an angle no larger than 80° with the floor. If Shemar places the ladder 2 feet from the wall and the ladder is 28 feet long, what angle will the ladder make with the floor? Will this meet the safety recommendation? If not, where should he place the foot of the ladder to conform to the safety recommendation?

Solution: The wall meets the floor at a right angle, and the ladder forms the hypotenuse of the right triangle. The side *adjacent to the angle* the ladder makes with the floor is distance between the wall and the foot of the ladder.

Figure 8.28

$$\cos \angle X = \frac{\text{adjacent}}{\text{hypotenuse}} = \frac{2}{28} \approx 0.0714$$. The angle whose cosine is 0.0714

is $X = \cos^{-1}(0.0714) \approx 85.9°$. This is larger than the recommended maximum 80° angle, so it does not meet the safety recommendation. If Shemar wants an angle of 80°, he can calculate the distance between the ladder and the wall.

$$\cos 80° = \frac{d}{28} \text{ so } 0.1736 = \frac{d}{28} \text{ and } d \approx 4.86 \text{ feet.}$$

Figure 8.29

The tangent of an angle is the ratio of the side opposite the angle to the side adjacent to the angle. The tangent of $\angle A$ is abbreviated $\tan\angle A$ or $\tan A$. Notice that:

$$\tan \angle A = \frac{\text{opposite}}{\text{adjacent}} = \frac{^{\text{opposite}}/_{\text{hypotenuse}}}{^{\text{adjacent}}/_{\text{hypotenuse}}} = \frac{\sin \angle A}{\cos \angle A}$$

Example 7

From his current position at sea, Sanjay can sight a 35° angle to the top of the lighthouse. He knows that the top of the lighthouse is 120 feet above the surface of the water. How far from the base of the lighthouse is Sanjay's boat?

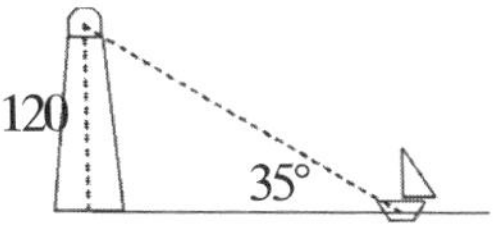

Solution: Using the right triangle formed by the lighthouse, the water surface, and the line of sight, you can use a tangent ratio to find the distance. The lighthouse is the side opposite the 35° angle,

120

35°

Figure 8.30

and the distance is the adjacent side. $\tan 35° = \dfrac{\text{opposite}}{\text{adjacent}} = \dfrac{120}{x}$.

Because $\tan 35° \approx 0.7002$, the equation becomes $0.7002 = \dfrac{120}{x}$. Cross-multiplying gives you $0.7002x = 120$, and $x = \dfrac{120}{0.7002} \approx 171.38$. Sanjay's boat is approximately 171.38 feet from the base of the lighthouse.

There are several mnemonics, or memory devices, to help you remember the trig ratios and their names. The most common one is *soh-cah-toa*, an invented word formed from the initial letters of *sine, opposite, hypotenuse, cosine, adjacent, hypotenuse, tangent, opposite, adjacent.* Of course, *soh-cah-toa* is not a real word, so it might be difficult to remember how to spell it, unless you already know the trig ratios. The mnemonic most students can remember easily is "oh heck, another hour of algebra."

$$\dfrac{\text{Oh}}{\text{Heck}} \quad \sin \quad \dfrac{\text{opposite}}{\text{hypotenuse}}$$

$$\dfrac{\text{Another}}{\text{Hour}} \quad \cos \quad \dfrac{\text{adjacent}}{\text{hypotenuse}}$$

$$\dfrac{\text{Of}}{\text{Algebra}} \quad \tan \quad \dfrac{\text{opposite}}{\text{adjacent}}$$

Figure 8.31

In addition to the definitions of the three major trig ratios, you should also commit a few key relationships to memory. These statements, called identities, are true for all angles. Two have already been mentioned: $\sin \angle A = \cos(90° - \angle A)$ and $\tan \angle A = \frac{\sin \angle A}{\cos \angle A}$. The third is called the **Pythagorean Identity.** You know that in any right triangle with sides a, b, and c, $a^2 + b^2 = c^2$. Because side a is the side opposite $\angle A$, side b is the side adjacent to $\angle A$, and c is the hypotenuse, you could also state the Pythagorean Theorem as (opposite)^2 + (adjacent)^2 = (hypotenuse)^2. Divide through by (hypotenuse)^2 and you produce the Pythagorean Identity.

$$(\text{opposite})^2 + (\text{adjacent})^2 = (\text{hypotenuse})^2$$

$$\dfrac{(\text{opposite})^2}{(\text{hypotenuse})^2} + \dfrac{(\text{adjacent})^2}{(\text{hypotenuse})^2} = \dfrac{(\text{hypotenuse})^2}{(\text{hypotenuse})^2}$$

$$\left(\dfrac{\text{opposite}}{\text{hypotenuse}}\right)^2 + \left(\dfrac{\text{adjacent}}{\text{hypotenuse}}\right)^2 = 1$$

$$(\sin \angle A)^2 + (\cos \angle A)^2 = 1$$

$$\sin^2 \angle A + \cos^2 \angle A = 1$$

Note that $\sin^2 \angle A$ and $\cos^2 \angle A$ are the traditional ways of writing (sin $\angle A)^2$ and $(\cos \angle A)^2$.

Example 8

In right triangle $\triangle RST$ with right angle $\angle S$, $\sin \angle R = \dfrac{5}{8}$. Find cos $\angle R$, tan $\angle R$, cos $\angle T$, sin $\angle T$, and tan $\angle T$.

Solution: Use the Pythagorean Identity to find cos $\angle R$.

$$\sin^2 \angle R + \cos^2 \angle R = 1$$

$$\left(\frac{5}{8}\right)^2 + \cos^2 \angle R = 1$$

$$\frac{25}{64} + \cos^2 \angle R = 1$$

$$\cos^2 \angle R = 1 - \frac{25}{64}$$

$$\cos^2 \angle R = \frac{39}{64}$$

$$\cos \angle R = \frac{\sqrt{39}}{8}$$

Once you know sin $\angle R$ and cos $\angle R$, you can find

$\tan \angle R = \dfrac{\sin \angle R}{\cos \angle R} = \dfrac{5/8}{\sqrt{39}/8} = \dfrac{5}{\sqrt{39}} = \dfrac{5\sqrt{39}}{39}$. The trig functions for $\angle T$

will be related: $\sin \angle T = \cos \angle R = \dfrac{\sqrt{39}}{8}$, and $\cos \angle T = \sin \angle R = \dfrac{5}{8}$.

Once again, you can find the tangent ratio by dividing the sine

ratio by the cosine ratio. $\tan \angle T = \dfrac{\sin \angle T}{\cos \angle T} = \dfrac{\sqrt{39}/8}{5/8} = \dfrac{\sqrt{39}}{5}$.

Lesson 8-3 Review

Use an appropriate trigonometric ratio to find the values of x and y in each diagram.

1.

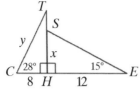

Figure 8.32

2.

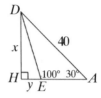

Figure 8.33

3. From a point 100 yards directly south of the western end of the lake, Chris measures an angle of 55° to sight the eastern end of the lake. How long is the lake?

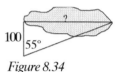

Figure 8.34

4. Jeff is designing a custom bike, and needs a support frame in the shape of a right triangle. The shorter leg must measure 14 inches and the hypotenuse 28 inches. Find the acute angles of the triangle to the nearest degree and the length of the longer leg to the nearest tenth of an inch.

5. In right triangle $\triangle ABC$ with right angle at B, $\cos \angle A = \frac{3}{5}$. Find sin $\angle A$, tan $\angle A$, sin $\angle C$, cos $\angle C$, and tan $\angle C$.

Lesson 8-4: Special Right Triangles

Trigonometry allows you to find missing sides or missing angles in any right triangle, assuming you have a little information and a trig table or a scientific calculator. There are, however, a few right triangles that appear so often that it's wise to know their proportions.

An isosceles triangle is a triangle with two sides of equal length. In an **isosceles right triangle**, the congruent sides will be the legs that form the right angle. If two sides of a triangle are congruent, the angles opposite those sides are congruent, so in an isosceles right triangle, the two acute angles have the same measure. You know that the acute angles are complementary, so each must be 45°. You'll sometimes hear an isosceles right triangle referred to as a **45°–45°–90° triangle**.

If you know the lengths of the legs of an isosceles right triangle, finding the length of the hypotenuse is just a matter of using the Pythagorean Theorem. If you do the calculation with the aid of a calculator, you may not notice a pattern immediately. If your triangle has legs of length 7, and you say $7^2 + 7^2 = 49 + 49 = 98$, and then use your calculator, you'll get an approximate value of 9.899. If, instead, you simplify the radical to get the exact value, you find that $\sqrt{98} = \sqrt{49 \cdot 2} = 7\sqrt{2}$. The length of the hypotenuse is the length of a side times the square root of 2.

Figure 8.35

In isosceles right triangles, the lengths of the legs are equal and the hypotenuse is $\sqrt{2}$ times a leg. When you start to compute trig ratios, you find $\sin 45° = \frac{s}{s\sqrt{2}} = \frac{1}{\sqrt{2}}$ and $\cos 45° = \frac{s}{s\sqrt{2}} = \frac{1}{\sqrt{2}}$. The sine and cosine of 45° are the same because the opposite and the adjacent sides are the same length; both equal $\frac{1}{\sqrt{2}}$, or, if you rationalize the denominator, $\frac{\sqrt{2}}{2}$. The tangent of 45° is the ratio of the two equal sides, so it is always 1. $\tan 45° = \frac{s}{s} = 1$.

Example 1

Find the length of the hypotenuse of an isosceles right triangle with a leg of length $5\sqrt{2}$.

Solution: The hypotenuse of an isosceles right triangle is equal to a side times $\sqrt{2}$. Because the side is $5\sqrt{2}$, the hypotenuse is $5\sqrt{2} \cdot \sqrt{2} = 5 \cdot 2 = 10$.

Example 2

A square has a diagonal that measures 9 cm. Find the area of the square.

Solution: To find the area, you need to know the length of a side. The diagonal divides the square into two isosceles right triangles, so the length of the diagonal, or hypotenuse, is $\sqrt{2}$ times the length of a side. That means $s\sqrt{2} = 9$, so $s = \frac{9}{\sqrt{2}}$. $A = s^2 = \left(\frac{9}{\sqrt{2}}\right)^2 = \frac{81}{2}$.

The area of the square is 40.5 cm².

The other special right triangle is the **30°–60°–90° right triangle**, but it's easier to understand where the relationships come from if you start with an equilateral triangle. An altitude drawn from any vertex of an equilateral triangle, to the opposite side, divides the equilateral triangle into two right triangles. $\angle ITP$ and $\angle ITN$ are right angles, because the altitude is perpendicular to the base, and perpendiculars form right angles. $\overline{PI} \cong \overline{IN}$ because $\triangle PIN$ is equilateral, and $\overline{IT}$ is equal to itself. So by the HL congruence theorem, $\triangle ITP \cong \triangle ITN$.

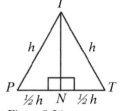

Figure 8.36

Because they are corresponding parts of the congruent triangles, $\overline{PT} \cong \overline{TN}$, so each of them must be half as long as $\overline{PN}$, and $\angle PIT \cong \angle NIT$, so each is half as large as $\angle PIN$; therefore, m$\angle PIT =$ m$\angle NIT = 30°$. Each of the congruent right triangles is a 30°–60°–90° right triangle. The side opposite the 30° angle is half as long as the hypotenuse. To find the length of $\overline{IT}$, use the Pythagorean Theorem.

$$a^2 + b^2 = c^2$$

$$\left(\frac{1}{2}h\right)^2 + b^2 = h^2$$

$$\frac{h^2}{4} + b^2 = h^2$$

$$b^2 = \frac{3h^2}{4}$$

$$b = \frac{h\sqrt{3}}{2}$$

The sides of a 30°–60°–90° right triangle are hypotenuse (opposite the right angle), half the hypotenuse (opposite the 30° angle), and half the hypotenuse times $\sqrt{3}$ (opposite the 60° angle).

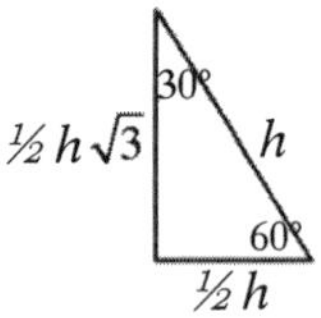

Because the opposite side for the 30° angle is the adjacent side for the 60° angle, and vice versa, the values for the sine and cosine of 30° and 60° seem to swap.

Figure 8.37

$$\sin 30° = \frac{\frac{1}{2}h}{h} = \frac{1}{2} \qquad \sin 60° = \frac{\frac{1}{2}h\sqrt{3}}{h} = \frac{\sqrt{3}}{2}$$

$$\cos 30° = \frac{\frac{1}{2}h\sqrt{3}}{h} = \frac{\sqrt{3}}{2} \qquad \cos 60° = \frac{\frac{1}{2}h}{h} = \frac{1}{2}$$

The tangent ratios for 30° and 60° are reciprocals.

$$\tan 30° = \frac{\frac{1}{2}h}{\frac{1}{2}h\sqrt{3}} = \frac{1}{\sqrt{3}} = \frac{\sqrt{3}}{3} \text{ and } \tan 60° = \frac{\frac{1}{2}h\sqrt{3}}{\frac{1}{2}h} = \sqrt{3}$$

Example 3

Find the lengths of the unknown sides in △XYZ.

Solution: △XYZ is a 30°– 60° – 90° right triangle, so the side opposite the 30° angle is half as long as the hypotenuse, and the side opposite the 60° angle is half the hypotenuse times the square root of three.

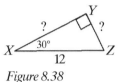
Figure 8.38

$$YZ = \frac{1}{2}(12) = 6, \text{ and } XY = 6\sqrt{3}.$$

Example 4

△QRA and △QUA are right triangles with $\overline{QR} \perp \overline{RA}$, $\overline{QU} \perp \overline{UA}$, and △QRA ≅ △QUA . m∠RQA = 30°, m∠UQA = 30°, and RA = 3 cm. Find QA and QU.

Figure 8.39

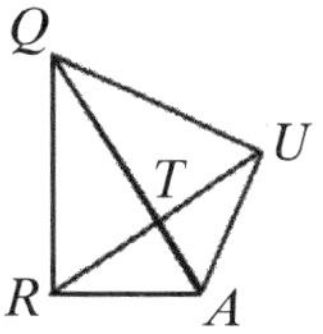

Solution: $\overline{QA}$ is the hypotenuse of 30°–60°–90° triangle △QRA. In △QRA, $\overline{RA}$, the side opposite the 30° angle, measures 3 cm. That should be half the hypotenuse, so QA = 6 cm. $\overline{QU}$ is the side opposite the 60° angle in △QUA. $\overline{QA}$ is also the hypotenuse of △QUA, and QA = 6 cm. △QRA ≅ △QUA, so UA = 3 cm. $\overline{QU}$ should be half the hypotenuse times the square root of three, so QU = 3√3 .

More Memory Work			
Angle	Sine	Cosine	Tangent
30°	$\frac{\sqrt{1}}{2}$	$\frac{\sqrt{3}}{2}$	$\frac{1}{\sqrt{3}}$
45°	$\frac{\sqrt{2}}{2}$	$\frac{\sqrt{2}}{2}$	1
60°	$\frac{\sqrt{3}}{2}$	$\frac{\sqrt{1}}{2}$	$\sqrt{3}$

Figure 8.40

Lesson 8-4 Review

Find the missing measurements in each figure.

1.

   Figure 8.41

2.

   Figure 8.42

3.

   Figure 8.43

Answer Key

Lesson 8-1

1. a. $m\angle A = 60°$ b. $m\angle Y = 7°$ c. $m\angle B = 43°$

2. a. $60^2 + 35^2 = 3600 + 1225 = 4825$ b. $AB^2 + 80^2 = 82^2$ c. $BO^2 + 4^2 = 7^2$

   $EA = \sqrt{4825} \approx 69.5$ $AB^2 + 6400 = 6724$ $BO^2 + 16 = 49$

                                           $AB^2 = 324$ $BO^2 = 33$

                                             $AB = 18$ $BO = \sqrt{33} \approx 5.7$

3. $\dfrac{ST}{EA} = \dfrac{TR}{AM}$ $\dfrac{ST}{EA} = \dfrac{SR}{EM}$

   $\dfrac{5}{15} = \dfrac{x}{36}$ $\dfrac{5}{15} = \dfrac{13}{y}$

   $15x = 180$ $5y = 195$

   $x = 12 = TR$ $y = 39 = EM$

4. $\triangle STC \sim \triangle SHG$ are similar right triangles, so their corresponding sides are in

   proportion. Write the proportion $\dfrac{ST}{SH} = \dfrac{CT}{GH}$ but be alert to the fact that SH is

   not given. Instead, you are told the values of HT and ST.

   $SH = HT + ST = 18 + 12 = 30$.

   Substitute in the proportion $\dfrac{12}{30} = \dfrac{10}{GH}$ and solve.

   $12 \cdot GH = 30 \cdot 10 \Rightarrow GH = \dfrac{300}{12} = 25$

   Marisa's house is 25 feet tall.

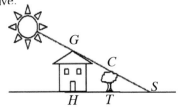

Figure 8.44

Lesson 8-2

1. $\dfrac{RU}{SU} = \dfrac{SU}{UT}$

   $\dfrac{9}{15} = \dfrac{15}{x}$

   $9x = 225$

   $x = 25 = UT$

2. $\dfrac{4}{a} = \dfrac{a}{8}$ $\qquad$ $\dfrac{4}{b} = \dfrac{b}{8}$

   $a^2 = 32$ $\qquad\quad$ $b^2 = 32$

   $a = 4\sqrt{2}$ $\qquad$ $b = 4\sqrt{2}$

3. $\dfrac{5}{7} = \dfrac{7}{y}$ $\qquad\qquad$ $\dfrac{5}{x} = \dfrac{x}{5 + 9.8}$

   $5y = 49$ $\qquad\qquad$ $x^2 = 5(14.8) = 74$

   $y = 9.8 = DC$ $\qquad$ $x = \sqrt{74} \approx 8.6 = AB$

Lesson 8-3

1. $\tan 15° = \dfrac{x}{12}$ $\qquad\qquad\qquad$ $\cos 28° = \dfrac{8}{y}$

   $x \approx 12(0.2679) \approx 3.2$

   $\qquad\qquad\qquad\qquad\qquad$ $y \approx \dfrac{8}{0.8829} \approx 9.1$

2. $\sin 30° = \dfrac{x}{40}$ $\qquad\qquad$ $\tan 80° = \dfrac{x}{y} = \dfrac{20}{y}$

   $x = 40\left(\tfrac{1}{2}\right) = 20$

   $\qquad\qquad\qquad\qquad$ $y \approx \dfrac{20}{5.6713} \approx 3.5$

3. $\tan 55° = \dfrac{x}{100}$

   $x \approx 100(1.4281) \approx 142.81$

   The lake is approximately 142.81 yards long.

4. Call the smaller angle $\angle A$. $\sin \angle A = \dfrac{14}{28} = \dfrac{1}{2}$. $\angle A = \sin^{-1}\left(\dfrac{1}{2}\right) = 30°$.

The second acute angle is $60°$. The longer leg can be found by

$$\cos 30° = \dfrac{x}{28}$$

$$x \approx 28(0.8660) \approx 24.2$$

The longer leg is approximately 24.2 inches.

5. $\sin^2 \angle A + \cos^2 \angle A = 1$ $\tan \angle A = \dfrac{\sin \angle A}{\cos \angle A} = \dfrac{\frac{4}{5}}{\frac{3}{5}} = \dfrac{4}{3}$

$(\sin \angle A)^2 + \left(\dfrac{3}{5}\right)^2 = 1$

$(\sin \angle A)^2 = 1 - \dfrac{9}{25} = \dfrac{16}{25}$ $\sin \angle C = \cos \angle A = \dfrac{3}{5}$

$\cos \angle C = \sin \angle A = \dfrac{4}{5}$

$\sin \angle A = \dfrac{4}{5}$

$\tan \angle C = \dfrac{\sin \angle C}{\cos \angle C} = \dfrac{\frac{3}{5}}{\frac{4}{5}} = \dfrac{3}{4}$

Lesson 8-4

1. $x = 7\sqrt{2}$

2. $x = \frac{1}{2}(12) = 6$, $y = \frac{1}{2}(12)\sqrt{3} = 6\sqrt{3}$

3. $x\sqrt{2} = 5\sqrt{2} \Rightarrow x = 5 \cdot$

$y = \frac{1}{2}w$, $x = \frac{1}{2}w\sqrt{3} = y\sqrt{3}$, so $5 = y\sqrt{3} \Rightarrow y = \dfrac{5}{\sqrt{3}} = \dfrac{5\sqrt{3}}{3} \approx 2.8867$.

$x = y + z$ so $5 = \dfrac{5\sqrt{3}}{3} + z \Rightarrow z = \dfrac{15 - 5\sqrt{3}}{3} \approx 2.1132$.

$y = \frac{1}{2}w \Rightarrow w = 2y$ so $w = 2\left(\dfrac{5\sqrt{3}}{3}\right) = \dfrac{10\sqrt{3}}{3} \approx 2(2.8867) \approx 5.7734$

Polygons

Lesson 9-1: Polygons

The term **polygon** is used to describe a closed figure made up of line segments that intersect only at their endpoints. *ABCDE* is not a polygon, because $\overline{BC}$ intersects $\overline{AE}$ at a point that is not an endpoint. The endpoints of the segments are called the vertices of the polygon and the line segments are its sides.

Figure 9.1

Polygons are often classified by the number of their sides. Some of the common names are given here. If there is no special name for a polygon, it can simply be called an *n*-gon. (For example, one with 13 sides would be called a 13-gon.)

- ▶ Triangle—3 sides
- ▶ Quadrilateral—4 sides
- ▶ Pentagon—5 sides
- ▶ Hexagon—6 sides

- ▶ Heptagon—7 sides
- ▶ Octagon—8 sides
- ▶ Nonagon—9 sides
- ▶ Decagon—10 sides

Example 1

Classify each polygon by the number of sides.

a.    b.   c.   d.

Figure 9.2

Solution:

a. 10 sides—decagon   c. 7 sides— heptagon

b. 8 sides— octagon   d. 6 sides— hexagon

Figure 9.3.

A line segment that connects two non-adjacent vertices—that is, a line segment that is not itself a side but has vertices of the polygon as its endpoints—is a diagonal. The number of diagonals in a polygon depends on the number of vertices.

There are *n* vertices from which to start drawing. Whichever vertex you choose, you can't draw to the starting vertex, or to either of the vertices adjacent to your starting point, so there are *n* − 3 vertices to which you can draw diagonals. You need to cut that in half, however, because the diagonal from *A* to *E* and the diagonal from *E* to *A*, for example, are actually the same line segment, and shouldn't be counted separately. The number of diagonals you can draw is $\frac{n(n-3)}{2}$.

Number of vertices	Number of diagonals
3	0
4	2
5	5
6	9

Example 2

Find the number of diagonals in a decagon.

Solution: A decagon would have 10 sides and 10 vertices. From each vertex, there would be seven diagonals that could be drawn. To get rid of that duplication, you have to divide 70 by 2. There are 35 diagonals in a decagon.

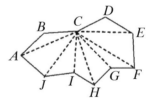

Figure 9.4

A polygon is **regular** if all its sides are congruent and all its angles are congruent. An equilateral triangle is regular, because all three of its sides are the same length and all its angles measure 60°. Because all its angles are right angles and all its side are congruent, a square is a regular polygon.

Example 3

Determine whether each polygon is regular.

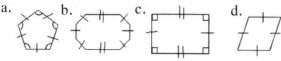

Figure 9.5

Solution:

a. Regular. Markings on the figure tell you that all sides are congruent and all angles are congruent.

b. Not regular. Only six of the eight sides of the octagon are the same length.

c. Not regular. Even though all its angles are congruent, its sides are not all the same length.

d. Regular. It is equilateral but its angles are not all the same size. The square is the only regular quadrilateral.

Lesson 9-1 Review

1. For each figure, decide whether the figure is a polygon. If not, explain why. If it is a polygon, name the polygon according to the number of sides and tell whether the polygon is regular.

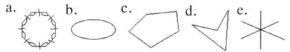

Figure 9.6

2. Find the number of diagonals in a polygon with:
   a. 15 sides
   b. 20 sides

Lesson 9-2: Trapezoids

A quadrilateral, in which one pair of opposite sides is parallel, is a **trapezoid**. The two non-parallel sides may meet the parallel sides at the same angle or at different angles. The parallel sides are the bases, and the non-parallel sides are called legs.

Because a trapezoid is formed by a pair of parallel line segments and two transversals, consecutive
angles between the parallels are supplementary. In
the figure, $\angle 1$ and $\angle 2$ are supplementary, and $\angle 3$
and $\angle 4$ are supplementary.

Figure 9.7

Example 1

If $RSTU$ is a trapezoid with $\overline{RS} \parallel \overline{TU}$, find the
measure of $\angle RST$.

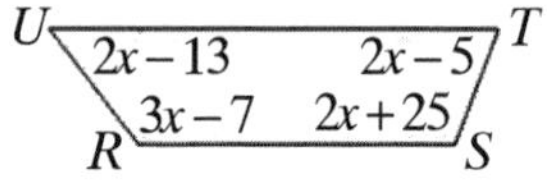

Figure 9.8

Solution: Use the fact that $\angle R$ and $\angle U$ are
supplementary (or that $\angle S$ and $\angle T$ are supplementary) to find
the value of x.

$m\angle R + m\angle U = 180°$

$3x - 7 + 2x - 13 = 180$

$5x - 20 = 180$

$5x = 200$

$x = 40$

$m\angle RST = 2x + 25 = 2(40) + 25 = 105°$

If the non-parallel sides are congruent, the trapezoid is an **isosceles
trapezoid**, and the base angles are congruent. Suppose $ABCD$ is a trapezoid with $AC = BD$. Extend sides $\overline{AC}$ and $\overline{BD}$ until
they meet at point E. In $\triangle ABE$, $\overline{CD} \parallel \overline{AB}$ divides sides
$\overline{AE}$ and $\overline{BE}$ proportionally. So $\frac{CE}{AC} = \frac{DE}{BD}$, but because $AC = BD$, $\frac{CE}{BD} = \frac{DE}{BD}$. Therefore $CE = DE$, and
$\triangle CDE$ is isosceles. $\triangle CDE \sim \triangle ABE$, so both triangles

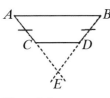

Figure 9.9

are isosceles. Base angles of an isosceles triangle are congruent, so $\angle A \cong$
$\angle B$. Once you know one pair of base angles are congruent, you can use
the supplementary angles to show that $\angle ACD \cong \angle BDC$.

Example 2

In trapezoid $ABCD$, $AD = BC$. Find the measure of $\angle B$.

Solution: Because the trapezoid is isosceles, m∠A = m∠B, so m∠B = $15x + 17$. Use the fact that ∠B and ∠C are supplementary to write and solve an equation to find the value of x.

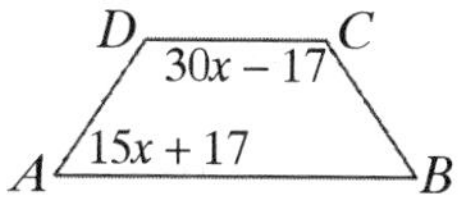

$15x + 17 + 30x - 17 = 180$

$45x = 180$

$x = 4$

m∠B = $15x + 17 = 15(4) + 17 = 77°$

Figure 9.10

As all quadrilaterals do, trapezoids have two diagonals. In most trapezoids, the diagonals are different lengths, but in an isosceles trapezoid, you can prove that the diagonals are congruent.

Given: Isosceles trapezoid $WXYZ$ with $\overline{ZW} = \overline{XY}$ shown in Figure 9.11

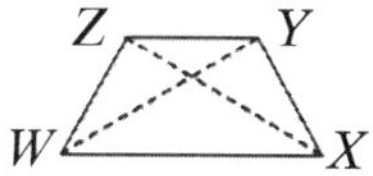

Prove: $\overline{WY} = \overline{XZ}$

Plan: Prove $\triangle WXY \cong \triangle XWZ$ by SAS, then use CPCTC to prove $\overline{WY} = \overline{XZ}$

Figure 9.11

Statement	Reason
1. Isosceles trapezoid $WXYZ$ with $\overline{ZW} = \overline{XY}$ (S)	1. Given
2. ∠$ZWX \cong$ ∠YXW (A)	2. Base angles of an isosceles trapezoid are congruent.
3. $\overline{WX} \cong \overline{WX}$ (S)	3. Reflexive
4. $\triangle WXY \cong \triangle XWZ$	4. SAS
5. $\overline{WY} = \overline{XZ}$	5. CPCTC

Figure 9.12

Example 3

Given: $\overline{TR} \cong \overline{TS}$, and $\overline{PQ} \parallel \overline{RS}$

Prove: $\overline{PS} \cong \overline{QR}$

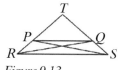

Figure 9.13

Solution: Show that $RSQP$ is an isosceles trapezoid by showing that $\overline{PQ} \parallel \overline{RS}$, which is given, and that $PR = QS$. Diagonals of an isosceles trapezoid are congruent.

Statement	Reason
1. $\overline{PQ} \parallel \overline{RS}$	1. Given
2. $RSPQ$ is a trapezoid.	2. A quadrilateral that has one pair of parallel sides is a trapezoid.
3. $\overline{TR} \cong \overline{TS}$	3. Given
4. $\angle TRS \cong \angle TSR$	4. If two sides of a triangle are congruent, the angles opposite those sides are congruent.
5. $RSPQ$ is an isosceles trapezoid.	5. If the base angles of a trapezoid are congruent, the trapezoid is isosceles.
6. $\overline{PS} \cong \overline{QR}$	6. Diagonals of an isosceles trapezoid are congruent.

Figure 9.14

The **median of a trapezoid** is the line segment that joins the midpoints of the non-parallel sides. In trapezoid $ABCD$, M is the midpoint of $\overline{AD}$ and N is the midpoint of $\overline{BC}$, so $\overline{MN}$ is the median of the trapezoid. The median of any trapezoid is parallel to the bases and its length is one-half the sum of the bases.

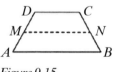

Figure 9.15

Extend the non-parallel sides until they meet at point E to form $\triangle AEB$ as shown in Figure 9.16 on page 171. Because $\overline{CD} \parallel \overline{AB}$, $\overline{CD}$ divides the sides of $\triangle AEB$ proportionally, and therefore $\frac{DE}{DA} = \frac{CE}{CB}$. M and N are the midpoints of $\overline{DA}$ and $\overline{CB}$, respectively, so $DA = 2DM$ and $CB = 2CN$. Substitute into the proportion for $\frac{DE}{2DM} = \frac{CE}{2CN}$ and multiply both sides by two, producing $\frac{DE}{DM} = \frac{CE}{CN}$. That means $\overline{CD}$ divides the sides of $\triangle MEN$ proportionally, so $\overline{CD} \parallel \overline{MN}$, and therefore, $\overline{MN} \parallel \overline{AB}$. The median is parallel to the bases.

To prove that the length of the median is one-half the sum of the bases, use the similar triangles. Because $\triangle MEN \sim \triangle AEB$, $\frac{MN}{AB} = \frac{ME}{AE}$, and $\triangle DEC \sim \triangle MEN$, $\frac{DC}{MN} = \frac{DE}{ME}$. Cross-multiplying each proportion, $MN \cdot AE = AB \cdot ME$ and $MN \cdot DE = DC \cdot ME$. Add the two equations and factor to get $MN(AE + DE) = ME(AB + DC)$.

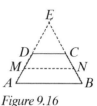

Figure 9.16

Break AE down into its parts.
$MN(AM + MD + DE + DE) = ME(AB + DC)$.
$AM = MD$, so substitute to get:

$$MN(MD + MD + DE + DE) = ME(AB + DC)$$
$$MN(2MD + 2DE) = ME(AB + DC)$$
$$2MN(MD + DE) = ME(AB + DC)$$
$$2MN \cdot ME = ME(AB + DC)$$

Divide to solve for MN, and you have:

$$MN = \frac{ME}{2ME}(AB + DC) = \frac{1}{2}(AB + DC).$$

The length of the median of any trapezoid is the average of the lengths of the bases.

Example 4

In trapezoid $ALGE$, $\overline{AL} \parallel \overline{EG}$, and $AL = 4x + 3$ and $EG = 3x + 2$. If $\overline{BR}$ is the median of the trapezoid, and $BR = 4x$, find the length of $\overline{BR}$.

Solution:

$$BR = \frac{1}{2}(AL + EG)$$

$$4x = \frac{1}{2}(4x + 3 + 3x + 2)$$

$$4x = \frac{1}{2}(7x + 5)$$

Multiplying both sides by 2 is easier than dealing with the fractions.
$8x = 7x + 5 \Rightarrow x = 5$. $BR = 4x = 4(5) = 20$.

1. Find the measure of $\angle B$ in each trapezoid.

a.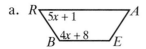

b.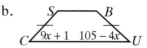

Figure 9.17

2. Find the length of *MA* in each trapezoid.

a.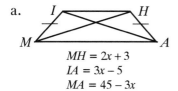

$MH = 2x + 3$
$IA = 3x - 5$
$MA = 45 - 3x$

b.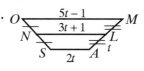

Figure 9.18

Lesson 9-3: Parallelograms

A quadrilateral in which both pairs of opposite sides are congruent is a **parallelogram**. Just as the trapezoid does, the parallelogram has consecutive angles that are supplementary.

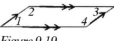

Figure 9.19

Example 1

In parallelogram $\square DUCK$, find the measure of $\angle C$.

Solution: $\angle D$ and $\angle K$ are supplementary, so

$$127 - 2x + 7x - 12 = 180$$
$$115 + 5x = 180$$
$$5x = 65$$
$$x = 13$$

$\angle C$ is supplementary to $\angle K$, and m$\angle K = 127 - 2(13) = 101°$, so $\angle C = 180 - 101 = 79°$.

If you draw one diagonal of a parallelogram, you can prove that it creates two congruent triangles.

Given: Parallelogram $\square LAMB$ with diagonal $\overline{LM}$ *Figure 9.21*

Prove: $\triangle LAM \cong \triangle MBA$

Plan: Use the parallel sides to find congruent angles. Prove the triangles congruent by ASA.

Statement	Reason
1. Parallelogram $\square LAMB$ with diagonal $\overline{LM}$	1. Given
2. $\overline{BM} \parallel \overline{LA}, \overline{BL} \parallel \overline{MA}$	2. Opposite sides of a parallelogram are parallel.
3. $\angle BML \cong \angle MLA$ (A)	3. When parallel lines $(\overline{BM} \parallel \overline{LA})$ are cut by a transversal $(\overline{LM})$, alternate interior angles are congruent.
4. $\overline{LM} \cong \overline{LM}$ (S)	4. Reflexive
5. $\angle BLM \cong \angle AML$ (A)	5. When parallel lines $(\overline{BL} \parallel \overline{MA})$ are cut by a transversal $(\overline{LM})$, alternate interior angles are congruent.
6. $\triangle LAM \cong \triangle MBA$	6. ASA

Figure 9.22

Once you have proved $\triangle LAM \cong \triangle MBA$, CPCTC assures you that $\overline{LA} \cong \overline{MB}$ and $\overline{BL} \cong \overline{AM}$, so opposite sides of a parallelogram are congruent. In addition, $\angle A \cong \angle B$, which means that one pair of opposite angles are congruent. You can prove the other pair of opposite angles congruent as well, either by drawing the other diagonal and repeating this proof, or by using the fact that consecutive angles are supplementary and that supplements of congruent angles are congruent. Either way, you find that opposite angles of a parallelogram are congruent.

Example 2

In parallelogram $\square GOAT$, $GO = 5x + 2$, $OA = x + 2$, and $AT = 7x - 4$. Find the length of $\overline{OA}$.

Solution: $\overline{GO}$ and $\overline{AT}$ are opposite sides of a parallelogram and therefore congruent. Write the equation $5x + 2 = 7x - 4$ and solve to find $-2x = -6 \Rightarrow x = 3$. Then $OA = 3 + 2 = 5$.

Example 3

In parallelogram $\square MICE$, $m\angle M = 3x + 11$ and $m\angle C = 4x - 6$. Find the measure of $\angle E$.

Solution: $\angle M$ and $m\angle C$ are opposite angles of a parallelogram and therefore congruent. Write the equation $3x + 11 = 4x - 6$ and solve $-x = -17 \Rightarrow x = 17$. Evaluating, you find that $m\angle M = 3(17) + 11 = 62$ and $m\angle C = 4(17) - 6 = 62$. Because $\angle E$ is supplementary to $\angle M$ or to $\angle C$, $m\angle E = 180 - 62 = 118°$

If you draw both diagonals of a parallelogram, you create four triangles, and you can show that there are two pairs of congruent triangles. In the figure, you can prove $\triangle SHE \cong \triangle ORE$ and $\triangle HOE \cong \triangle RSE$.

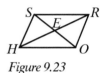

Figure 9.23

Given: Parallelogram $\square HORS$ with diagonals $\overline{HR}$ and $\overline{SO}$ intersecting at E

Prove: $\triangle SHE \cong \triangle ORE$ and $\triangle HOE \cong \triangle RSE$

Plan: Use the parallel lines to find congruent angles, and use the fact that opposite sides are congruent. Prove triangles congruent by ASA.

Statement	Reason
1. Parallelogram $\square HORS$ with diagonals $\overline{HR}$ and $\overline{SO}$ intersecting at E.	1. Given
2. $\overline{HO} \parallel \overline{SR}$, $\overline{HS} \parallel \overline{OR}$	2. Opposite sides of a parallelogram are parallel.
3. $\angle SHE \cong \angle ORE$ (A)	3. When parallel lines ($\overline{HS} \parallel \overline{OR}$) are cut by a transversal ($\overline{HR}$), alternate interior angles are congruent.
4. $\overline{HS} \cong \overline{RO}$ (S)	4. Opposite sides of a parallelogram are congruent.
5. $\angle HSE \cong \angle ROE$ (A)	5. When parallel lines ($\overline{HS} \parallel \overline{OR}$) are cut by a transversal ($\overline{SO}$), alternate interior angles are congruent.
6. $\triangle SHE \cong \triangle ORE$	6. ASA

(Figure 9.24 continued on page 175)

Statement (continued)	Reason (continued)
7. ∠HOE ≅ ∠RSE (A)	7. When parallel lines ($\overline{HO} \parallel \overline{SR}$) are cut by a transversal ($\overline{SO}$), alternate interior angles are congruent.
8. $\overline{HO} \cong \overline{RS}$ (S)	8. Opposite sides of a parallelogram are congruent.
9. ∠OHE ≅ ∠SRE (A)	9. When parallel lines ($\overline{HO} \parallel \overline{SR}$) are cut by a transversal ($\overline{HR}$), alternate interior angles are congruent.
10. △HOE ≅ △RSE	10. ASA

Figure 9.24

Once you prove that $\triangle SHE \cong \triangle ORE$ and $\triangle HOE \cong \triangle RSE$, you can show by CPCTC that $\overline{HE} \cong \overline{RE}$ and $\overline{SE} \cong \overline{OE}$. That tells you that the diagonals of a parallelogram bisect each other.

Example 4

In parallelogram $\square ZEBR$, diagonals $\overline{ZB}$ and $\overline{ER}$ intersect at A. If $ZA = 2x + 1$, $BA = 3x - 1$, and $EA = x + 1$, find the length of $\overline{ER}$.

Solution: In a parallelogram, the diagonals bisect each other, so $ZA = BA$. Write the equation $2x + 1 = 3x - 1$ and solve to find $x = 2$. To find the length of $\overline{ER}$, remember that EA is half of ER, so $ER = 2EA = 2(x + 1) = 2(2+1) = 6$.

Each of the examples you've looked at so far begins with the knowledge that the polygon is a parallelogram, but often you need to prove that a quadrilateral is a parallelogram. There are many different ways to do that. You can prove:

▶ Both pairs of opposite sides parallel.

▶ Both pairs of opposite sides congruent.

▶ Both pairs of opposite angles congruent.

▶ One pair of opposite sides both parallel and congruent.

▶ The diagonals bisect each other.

Proving any one of these statements is sufficient to guarantee that your quadrilateral is a parallelogram, but notice that each of the statements has two parts to it. You must prove two pairs of sides parallel, or two pairs of sides congruent, or two pairs of angles congruent. You must show one pair of sides has two properties: They are parallel and congruent. Or you must show that the long diagonal bisects the short one and the short one bisects the long one. Always be certain you've done both bits.

Subroutines

Both pairs of opposite sides parallel → parallelogram

$\overline{AB} \parallel \overline{DC}$

$\overline{AD} \parallel \overline{BC}$

ABCD is a parallelogram   A quadrilateral with two pair of parallel sides is a parallelogram.

Figure 9.25

Both pairs of opposite sides congruent → parallelogram

$\overline{AB} \cong \overline{DC}$

$\overline{AD} \cong \overline{BC}$

ABCD is a parallelogram   If both pairs of opposite sides are congruent, the quadrilateral is a parallelogram.

Figure 9.26

Both pairs of opposite angles congruent → parallelogram

$\angle A \cong \angle C$

$\angle B \cong \angle D$

ABCD is a parallelogram.   If both pairs of opposite angles are congruent, the quadrilateral is a parallelogram.

Figure 9.27

One pair of opposite sides both parallel and congruent → parallelogram

$\overline{AB} \parallel \overline{DC}$

$\overline{AB} \cong \overline{DC}$

ABCD is a parallelogram   If one pair of opposite sides is both parallel and congruent, the quadrilateral is a parallelogram.

Figure 9.28

Diagonals bisect each other → parallelogram

$AE \cong EC$	
BD bisects AC	
$BE \cong ED$	
AC bisects BD	
$ABCD$ is a parallelogram.	If the diagonals bisect each other, the quadrilateral is a parallelogram.

Figure 9.29

Example 5

Given: $\angle RNO$ is supplementary to $\angle NHI$, $\overline{NO} \cong \overline{IH}$, and $\angle NOR \cong \angle NRO$

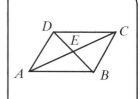

Prove: $RHIN$ is a parallelogram

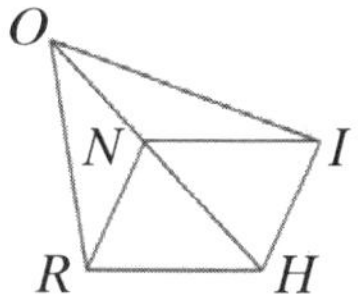

Solution: Work with the angles to show that

Figure 9.30

$\overline{RN} \parallel \overline{IH}$, and then with isosceles triangle $\triangle NOR$ to

prove $\overline{RN} \cong \overline{IH}$

Statement	Reason
1. $\angle RNO$ is supplementary to $\angle NHI$.	1. Given
2. $\angle RNO$ is supplementary to $\angle RNH$.	2. Linear pairs are supplementary.
3. $\angle NHI \cong \angle RNH$	3. Supplements of the same angle are congruent to each other.
4. $\overline{RN} \parallel \overline{IH}$	4. If alternate interior angles are congruent, lines are parallel.
5. $\overline{NO} \cong \overline{IH}$	5. Given
6. $\angle NOR \cong \angle NRO$	6. Given
7. $\overline{NO} \cong \overline{RN}$	7. If two angles of a triangle are congruent, the sides opposite those angles are congruent.
8. $\overline{RN} \cong \overline{IH}$	8. Transitivity
9. $RHIN$ is a parallelogram.	9. If one pair of opposite sides ($\overline{RN}, \overline{IH}$) is parallel (#4) and congruent (#8), the quadrilateral is a parallelogram.

Figure 9.31

Example 6

Given: $\overline{TR}$ is the midsegment of $\triangle HOG$,
$GT = HA$, and $OR = HW$

Prove: $WART$ is a parallelogram

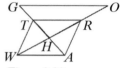

Figure 9.32

Solution: Work with $\triangle HOG$ and show that
$GT = TH$ and $OR = RH$, then use transitivity to show that the
diagonals of $WART$ bisect each other.

Statement	Reason
1. TR is the midsegment of $\triangle HOG$.	1. Given
2. T is the midpoint of GH and R is the midpoint of HO.	2. The midsegment of a triangle connects the midpoints of two sides.
3. $GT = TH$ and $OR = RH$	3. A midpoint divides a segment into two segments of equal length.
4. $GT = HA$ and $OR = HW$	4. Given
5. $TH = HA$ and $RH = HW$	5. Transitivity
6. WR and AT bisect each other.	6. A bisector divides a segment into two segments of equal length.
7. $WART$ is a parallelogram.	7. If the diagonals of a quadrilateral bisect each other, the quadrilateral is a parallelogram.

Figure 9.33

Lesson 9-3 Review

1. In parallelogram $\square STAG$, m$\angle S = 5x - 9$ and m$\angle T = 165 - 2x$. Find the measure of $\angle A$.

2. In parallelogram $\square HART$, $HA = 2x - 12$, $AR = x + 9$, and $RT = x - 6$. Find the length of $\overline{HT}$.

3. In parallelogram $\square BUCK$, m$\angle U = 5x + 13$ and m$\angle K = 7x - 21$. Find the measure of $\angle B$.

4. In parallelogram $\square WHAL$, diagonals $\overline{WA}$ and $\overline{HL}$ intersect at E. If $WE = x + 3$, $AE = 2x - 5$, and $HL = 3x - 7$, find the length of $\overline{HE}$.

5. □*MNKY* is a parallelogram, ∠*M* ≅ ∠*YOM*, ∠*K* ≅ ∠*NEK*, *E* is the midpoint of $\overline{YK}$, and *O* is the midpoint of $\overline{MN}$. Prove that *ONEY* is a parallelogram.

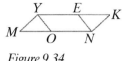

Figure 9.34

Lesson 9-4: Rhombuses, Rectangles, and Squares

Within the family of parallelograms, there are groups of parallelograms with specific properties that set them apart. Each of them is a parallelogram, and has all the properties of a parallelogram, but they have additional, special characteristics.

A **rhombus** is a parallelogram in which all four sides are congruent. It does not have to have right angles. Every square is a rhombus, but not every rhombus is a square.

Rhombi (or rhombuses) are parallelograms, with all the properties of parallelograms. Their opposite sides are parallel and congruent, their opposite angles are congruent, their consecutive angles are supplementary, and their diagonals bisect each other.

It is easy to prove that the four triangles formed by drawing the diagonals of a rhombus are all congruent, and then it's possible to show that the diagonals are also perpendicular.

Given: Rhombus □*ABCD* with diagonals $\overline{AC}$ and $\overline{BD}$

Prove: $\overline{AC} \perp \overline{BD}$

Figure 9.35

Statement	Reason
1. Rhombus □*ABCD* with diagonals $\overline{AC}$ and $\overline{BD}$.	1. Given
2. Rhombus □*ABCD* is a parallelogram.	2. A rhombus is a parallelogram.
3. $\overline{AE} \cong \overline{EC}$ and $\overline{BE} \cong \overline{ED}$	3. The diagonals of a parallelogram bisect each other.
4. $\overline{AB} \cong \overline{BC} \cong \overline{CD} \cong \overline{AD}$	4. A rhombus is a parallelogram with four congruent sides.
5. △*AEB* ≅ △*CEB* ≅ △*AED* ≅ △*CED*	5. SSS
6. ∠*AEB* ≅ ∠*CEB* ≅ ∠*AED* ≅ ∠*CED*	6. CPCTC

(Figure 9.36 continued on page 180)

Statement (continued)	Reason (continued)
7. $\angle AEB$ is supplementary to $\angle CEB$	7. Linear pairs are supplementary.
8. $m\angle AEB + m\angle CEB = 180°$	8. Definition of supplementary
9. $m\angle AEB + m\angle AEB = 2\ m\angle AEB = 180°$	9. Substitution
10. $m\angle AEB = 90°$	10. Division
11. $m\angle AEB = m\angle CEB = m\angle AED = m\angle CED = 90°$	11. Substitution
12. $\angle AEB$, $\angle CEB$, $\angle AED$, and $\angle CED$ are right angles.	12. Right angles measure 90°
13. $\overline{AC} \perp \overline{BD}$	13. Perpendiculars form right angles.

Figure 9.36

Example 1

If $\square SPIN$ is a rhombus, $m\angle NSE = 2x - 4$ and $m\angle SNE = 6x - 2$, find the measure of $\angle PIE$.

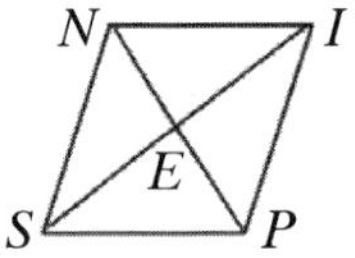

Solution: Because $\square SPIN$ is a rhombus, $\overline{NP} \perp \overline{SI}$, so $\angle NES$ is a right angle. $\angle NSE$, $\angle SNE$, and $\angle NES$ are the three angles of $\triangle NES$, so they add to 180°.

Figure 9.37

$$m\angle NSE + m\angle SNE + m\angle NES = 180°$$
$$(2x - 4) + (6x - 2) + 90 = 180$$
$$8x - 6 = 90$$
$$8x = 96$$
$$x = 12$$

$\angle PIE \cong \angle NSE$, so $m\angle PIE = 2x - 4 = 2(12) - 4 = 20°$

A **rectangle** is a parallelogram in which all angles are right angles. The rectangle is a parallelogram, so it has all the properties of a parallelogram. Its opposite sides are congruent and parallel, its opposite angles are congruent, and its consecutive angles are supplementary.

As in any parallelogram, the diagonals of a rectangle bisect each other. It is possible to show that the diagonals of a rectangle are congruent.

Drawing the two diagonals in a rectangle such as □*PAGE* creates two overlapping right triangles (△*APE* and △*PAG*). It may be easier to see them if you redraw them separately.

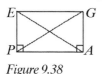
Figure 9.38

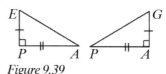
Figure 9.39

Because □*PAGE* is a parallelogram, opposite sides are congruent, so $\overline{EP} \cong \overline{GA}$. ∠*EPA* and ∠*GAP* are right angles, and all right angles are congruent, so ∠*EPA* ≅ ∠*GAP*. Segment $\overline{PA}$ is congruent to itself, so △*APE* ≅ △*PAG* by SAS. Then, by CPCTC, $\overline{EA} \cong \overline{PG}$, so the diagonals of the rectangle are congruent.

Example 2

In rectangle □*TYPE*, *TP* = 12*x* + 6, *YE* = 14*x* − 8, and *TY* = 10*x* + 2. Find the lengths of $\overline{PE}$ and $\overline{TE}$.

Solution: The diagonals of rectangle □*TYPE* are congruent, allowing you to write the equation 12*x* + 6 = 14*x* − 8. Solving 12*x* + 6 = 14*x* − 8 gives 2*x* = 14 ⇒ *x* = 7. Opposite sides are congruent, so *PE* = *TY* = 10*x* + 2 = 10(7) + 2 = 72. To find the length of $\overline{TE}$, take advantage of the fact that △*TYE* is a right triangle, and use the Pythagorean Theorem. Leg *TY* = 72 and hypotenuse *YE* = 14(7) − 8 = 90, so

$$a^2 + b^2 = c^2$$
$$72^2 + b^2 = 90^2$$
$$5184 + b^2 = 8100$$
$$b^2 = 2916$$
$$b = 54$$

The length of $\overline{PE}$ is 72 and the length of $\overline{TE}$ is 54.

A parallelogram with four congruent sides and four right angles is a **square**. A square is both a rhombus and a rectangle, and so it has all the properties of a parallelogram, all the properties of a rhombus, and all the properties of a rectangle.

To prove that a quadrilateral is a rhombus, you must prove that it is a parallelogram, and then prove that adjacent sides are congruent. It's enough to prove that adjacent sides are congruent, because you know that in a parallelogram opposite sides are congruent. You can also prove that a quadrilateral is a rhombus by showing that it is a parallelogram and that the diagonals are perpendicular.

To prove that a quadrilateral is a rectangle, you must prove that it is a parallelogram, and then prove that at least one of its angles is a right angle. You only need to prove that one angle is a right angle, because in a parallelogram opposite angles are congruent and consecutive angles are supplementary. You can also prove that a quadrilateral is a rectangle by proving that it is a parallelogram and then proving its diagonals are congruent. To prove that a quadrilateral is a square, prove that it is both a rectangle and a rhombus.

Subroutines

Proving quadrilateral is a rhombus

$ABCD$ is a parallelogram.		or	$ABCD$ is a parallelogram.	
$\overline{AD} \cong \overline{BC}$			$\overline{AC} \perp \overline{BC}$	
$ABCD$ is a rhombus.	A parallelogram with four congruent sides is a rhombus.		$ABCD$ is a rhombus.	A parallelogram in which the diagonals are perpendicular is a rhombus.

Figure 9.40

Proving quadrilateral is a rectangle

$ABCD$ is a parallelogram.		or	$ABCD$ is a parallelogram.	
$\angle A$ is a right angle.			$AC = BD$	
$ABCD$ is a rectangle.	A parallelogram that contains a right angle is a rectangle.		$ABCD$ is a rectangle.	A parallelogram in which the diagonals are congruent is a rectangle.

Figure 9.41

Proving quadrilateral is a square

ABCD is a parallelogram.

ABCD is a rhombus.

ABCD is a rectangle.

ABCD is a square.   A parallelogram that is both a rhombus and a rectangle is a square.

Figure 9.42

Example 3

Given: $\overline{FL} \parallel \overline{AE}$, $\overline{FL} \cong \overline{AE}$, and $\overline{FE}$ bisects $\angle LEA$

Prove: *LEAF* is a rhombus

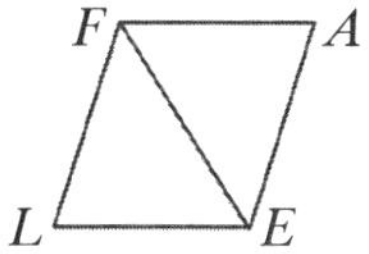

Solution:

Figure 9.43

Statement	Reason
1. $\overline{FL} \parallel \overline{AE}$	1. Given
2. $\overline{FL} \cong \overline{AE}$	2. Given
3. *LEAF* is a parallelogram.	3. If one pair of sides is both parallel and congruent, the quadrilateral is a parallelogram.
4. $\angle LFE \cong \angle AEF$	4. If parallel lines are cut by a transversal, alternate interior angles are congruent.
5. $\overline{FE}$ bisects $\angle LEA$	5. Given
6. $\angle LEF \cong \angle AEF$	6. A bisector divides the angle into two congruent angles.
7. $\angle LFE \cong \angle LEF$	7. Transitivity
8. $\overline{FL} \cong \overline{LE}$	8. If two angles of a triangle are congruent, the sides opposite those angles are congruent.
9. *LEAF* is a rhombus.	9. If adjacent sides of a parallelogram are congruent, the parallelogram is a rhombus.

Figure 9.44

Example 4

Given: $\overline{ER} \cong \overline{VR}$, $\overline{CE} \perp \overline{EV}$, and $\overline{OV} \perp \overline{EV}$

Prove: $COVE$ is a rectangle

Solution:

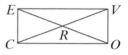

Figure 9.45

Statement	Reason
1. $\overline{ER} \cong \overline{VR}$	1. Given
2. $\angle REV \cong \angle RVE$	2. If two sides of a triangle are congruent, the angles opposite those angles are congruent.
3. $\overline{EV} \cong \overline{EV}$ (S)	3. Reflexive
4. $\overline{CE} \perp \overline{EV}$, $\overline{OV} \perp \overline{EV}$	4. Given
5. $\angle CEV$ is a right angle, $\angle OVE$ is a right angle	5. Perpendiculars form right angles.
6. $\angle CEV \cong \angle OVE$ (A)	6. All right angles are congruent.
7. $\triangle CVE \cong \triangle OEV$	7. ASA
8. $\overline{CE} \cong \overline{OV}$	8. CPCTC
9. $\overline{CE} \parallel \overline{OV}$	9. Two lines perpendicular to the same line are parallel to each other.
10. $COVE$ is a parallelogram.	10. If one pair of sides is both parallel and congruent, the quadrilateral is a parallelogram.
11. $\overline{CV} \cong \overline{OE}$	11. CPCTC
12. $COVE$ is a rectangle.	12. If the diagonals of a parallelogram are congruent, the parallelogram is a rectangle.

Figure 9.46

Example 5

Given: $\square MRGN$ is a parallelogram, $\overline{NA} \perp \overline{MR}$, $\overline{RI} \perp \overline{NG}$, and $\angle ANR \cong \angle ARN$

Prove: $ARIN$ is a square

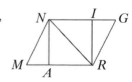

Figure 9.47

Solution:

Statement	Reason
1. $\overline{NA} \perp \overline{MR}$	1. Given
2. $\angle MAN$ is a right angle.	2. Perpendiculars form right angles.
3. $\overline{RI} \perp \overline{NG}$	3. Given
4. $\angle RIG$ is a right angle.	4. Perpendiculars form right angles.
5. $\angle MAN \cong \angle RIG$ (A)	5. All right angles are congruent.
6. $\square MRGN$ is a parallelogram.	6. Given
7. $\angle M \cong \angle G$ (A)	7. Opposite angles of a parallelogram are congruent.
8. $\overline{MN} \cong \overline{RG}$ (S)	8. Opposite sides of a parallelogram are congruent.
9. $\triangle MAN \cong \triangle GIR$	9. AAS
10. $\overline{NA} \cong \overline{IR}$	10. CPCTC
11. $\overline{MR} \parallel \overline{NG}$	11. Opposite sides of a parallelogram are parallel.
12. $\overline{NA} \parallel \overline{IR}$	12. Two lines perpendicular to parallel lines are parallel.
13. $ARIN$ is a parallelogram.	13. If one pair of sides is both parallel and congruent, the quadrilateral is a parallelogram.
14. $\angle NAR$ is a right angle.	14. Perpendiculars form right angles.
15. $ARIN$ is a rectangle.	15. A parallelogram with one right angle is a rectangle.
16. $\angle ANR \cong \angle ARN$	16. Given
17. $\overline{NA} \cong \overline{AR}$	17. If two angles of a triangle are congruent, the sides opposite those angles are congruent.
18. $ARIN$ is a rhombus.	18. If adjacent sides of a parallelogram are congruent, the parallelogram is a rhombus.
19. $ARIN$ is a square.	19. A parallelogram that is both a rhombus and a rectangle is a square.

Figure 9.48

Lesson 9-4 Review

1. $\square CARE$ is a rhombus, with diagonals $\overline{CR}$ and $\overline{EA}$ intersecting at T. If $m\angle TER = 8x + 3$ and $m\angle ERT = 2x - 3$, find the measure of $\angle TAR$.

2. In rectangle $\square FONT$, $FN = x + 6$, $TO = 2x + 1$, and $ON = x - 2$. Find the length of $\overline{ON}$.

Use Figure 9.49 for questions 3 and 4.

3. **Given:** $\overline{UP} \parallel \overline{TI}$, $\overline{UT} \parallel \overline{PC}$, $\overline{UE} \cong \overline{TP}$

   Prove: $PETU$ is a rectangle

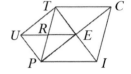

Figure 9.49

4. **Given:** $PETU$ is a rectangle, $\overline{UE} \cong \overline{CI}$, $\overline{TC} \cong \overline{PI}$

   Prove: $PICT$ is a rhombus

Lesson 9-5: Interior and Exterior Angles

An exterior angle of a triangle is equal to the sum of the two remote interior angles. Quadrilaterals and other polygons also have exterior angles, as well as interior angles, and, although the rules for their measures aren't quite so simple, there are patterns.

As with the three angles of a triangle, other polygons have a specific total for the measures of their interior angles. To determine the total of the measures of the interior angles of any polygon, start by dividing the polygon into triangles. You can do this by choosing one vertex of the polygon and drawing diagonals to each of the other vertices. This breaks up most of the angles in the polygon, and the angles in this example have been numbered.

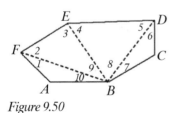

Figure 9.50

In this hexagon, four triangles were created by drawing diagonals. The total of the measures of the interior angles of this hexagon is the total of the measures of the angles in the four triangles formed by drawing the diagonals. In each triangle, the sum of the angles is 180°. If you add all the angles, you get

$$m\angle A + m\angle 1 + m\angle 10 + m\angle 2 + m\angle 3 + m\angle 9$$
$$+ \, m\angle 4 + m\angle 5 + m\angle 8 + m\angle 6 + m\angle C + m\angle 7 = 720°.$$

A little rearranging and renaming will give you the total of the measures of the angles of the hexagon.

$m\angle A + (m\angle 10 + m\angle 9 + m\angle 8 + m\angle 7) + m\angle C$
$+ (m\angle 5 + m\angle 6) + (m\angle 3 + m\angle 4) + (m\angle 1 + m\angle 2) = 720°$
$m\angle A + (m\angle B) + m\angle C + (m\angle D) + (m\angle E) + (m\angle F) = 720°$

The number of triangles produced by drawing the diagonals from one vertex is always two less than the number of sides. Each triangle contains 180 degrees, so the total of the measures of the angles of a polygon with n sides is $(n - 2)$ times 180.

Example 1

Find the total of the measures of the interior angles of a polygon with 13 sides.

Solution: A polygon with 13 sides will be divided into 11 triangles, each with 180°. The total of the measures of the interior angles is $11 \times 180 = 1,980°$.

If the polygon is regular, all the sides are the same length and all the angles are the same size, so the measure of any one of the angles is the total of all the angles divided by the number of angles.

The measure of an interior angle of a regular n-gon is $\dfrac{180(n-2)}{n}$.

Example 2

Find the measure of an interior angle of a regular decagon.

Solution: A decagon has 10 sides, and can be divided into eight triangles. The total of the measures of all 10 interior angles is $8 \times 180 = 1,440°$. The measure of each of the interior angles is

$\dfrac{180(10-2)}{10} = \dfrac{180 \times 8}{10} = \dfrac{1440}{10} = 144°$. The measure of an interior

angle of a regular decagon is 144°.

Choose a polygon and extend a side at each vertex, so that you create one exterior angle at each vertex. At each vertex, the exterior angle and the adjacent interior angle form a linear pair, totaling 180°. The polygon has n vertices, and at each of the n vertices, a pair of angles that add to 180°, so the n interior angles and n exterior angles together total 180n. The sum of the interior angles is $180(n - 2)$, so the sum of the exterior angles is $180n - 180(n - 2) = 360°$. No matter how many sides the polygon may have, the total of the exterior angles is 360°.

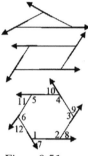

Figure 9.51

If the polygon is regular, the measure of each of the interior angles is the same and therefore the measure of each of the exterior angles the total, 360°, divided by the number of angles, or $\dfrac{360°}{n}$.

Example 3

Find the measure of an exterior angle of a regular octagon.

Solution: The total of the measures of the eight exterior angles is 360°, so the measure of one exterior angle of a regular octagon is

$$\frac{360°}{8} = 45°.$$

Lesson 9-5 Review

1. Find the total of the measures of the interior angles of a
   a. pentagon   b. octagon   c. 17-gon

2. Find the measure of an interior angle of
   a. a regular hexagon   b. a regular heptagon

3. Find the measure of an exterior angle of a regular decagon.

Answer Key

Lesson 9-1

1. a. Polygon, octagon, regular

   b. Not a polygon. The ellipse, or oval, is not made up of line segments.

   c. Polygon, pentagon, not regular

   d. Polygon, quadrilateral, not regular

   e. Not a polygon. The figure is not closed.

2. a. A polygon with 15 sides will have $\dfrac{15(15-3)}{2} = \dfrac{15(12)}{2} = \dfrac{180}{2} = 90$ diagonals

   b. A polygon with 20 sides will have $\dfrac{20(20-3)}{2} = \dfrac{20(17)}{2} = \dfrac{340}{2} = 170$ diagonals

Lesson 9-2

1. a. $m\angle R + m\angle B = 180°$
   $5x + 1 + 4x + 8 = 180$
   $9x + 9 = 180$
   $9x = 171$
   $x = 19$
   $m\angle B = 4x + 8 = 4(19) + 8 = 84°$

   b. $m\angle C = m\angle U$
   $9x + 1 = 105 - 4x$
   $13x = 104$
   $x = 8$
   $m\angle C = m\angle U = 73$
   $m\angle B = 180 - 73 = 107°$

2. a. $MH = IA$
   $2x + 3 = 3x - 5$
   $8 = x$
   $MA = 45 - 3x = 45 - 3(8) = 21$

   b. $NL = \frac{1}{2}(OM + SA)$
   $2NL = OM + SA$
   $2(3t + 1) = 5t - 1 + 2t$
   $6t + 2 = 7t - 1$
   $3 = t$
   $MA = 2AL = 2t = 6$

Lesson 9-3

1. $m\angle S + m\angle T = 180$
   $5x - 9 + 165 - 2x = 180$
   $3x + 156 = 180$
   $3x = 24$
   $x = 8$
   $m\angle T = 165 - 2(8) = 149°$
   $m\angle A = 180 - 149 = 31°$

2. $HA = RT$

   $2x - 12 = x - 6$

   $x = 6$

   $HT = AR = x + 9 = 6 + 9 = 15$

3. $m\angle U = m\angle K$

   $5x + 13 = 7x - 21$

   $34 = 2x$

   $17 = x$

   $m\angle U = m\angle K = 98$

   $m\angle B = 180 - 98 = 82°$

4. $WE = AE$

   $x + 3 = 2x - 5$

   $8 = x$

   $HE = \frac{1}{2}HL = \frac{1}{2}(3x - 7) = \frac{1}{2}(3 \cdot 8 - 7) = \frac{1}{2}(17) = 8.5$

5.

Statement	Reason
1. $\square MNKY$ is a parallelogram.	1. Given
2. $\angle M \cong \angle YOM$	2. Given
3. $\overline{MY} = \overline{YO}$	3. If two angles of a triangle are congruent, the sides opposite those angles are congruent.
4. $\angle K \cong \angle NEK$	4. Given
5. $\overline{EN} \cong \overline{NK}$	5. If two angles of a triangle are congruent, the sides opposite those angles are congruent.
6. $\overline{MY} \cong \overline{NK}$	6. Opposite sides of a parallelogram are congruent.
7. $\overline{YO} \cong \overline{EN}$	7. Transitivity
8. E is the midpoint of $\overline{YK}$, O is the midpoint of $\overline{MN}$.	8. Given
9. $YE = \frac{1}{2}YK$, $ON = \frac{1}{2}MN$	9. A midpoint divides a segment into two segments, each half as long as the original.
10. $\overline{MN} \cong \overline{KY}$	10. Opposite sides of a parallelogram are congruent.
11. $\frac{1}{2}MN = \frac{1}{2}YK$	11. Multiplication Property of Equality
12. $ON = YE$	12. Substitution
13. $ONEY$ is a parallelogram.	13. If two pairs of opposite sides are congruent, the quadrilateral is a parallelogram.

Figure 9.52

Lesson 9-4

1. $m\angle TER + m\angle ERT + m\angle ETR = 180$
    $m\angle TER + m\angle ERT = 90$
    $8x + 3 + 2x - 3 = 90$
    $10x = 90$
    $x = 9$
    $m\angle TAR = m\angle TER = 8x + 3 = 8(9) + 3 = 75°$

2. $FN = TO$
    $x + 6 = 2x + 1$
    $5 = x$
    $ON = 5 - 2 = 3$

3.

Statement	Reason
1. $\overline{UP} \parallel \overline{TI}, \overline{UT} \parallel \overline{PC}$	1. Given
2. *PETU* is a parallelogram.	2. If both pairs of opposite sides are parallel, the quadrilateral is a parallelogram.
3. $\overline{UE} \cong \overline{TP}$	3. Given
4. *PETU* is a rectangle.	4. If the diagonals of a parallelogram are congruent, the parallelogram is a rectangle.

Figure 9.53

4.

Statement	Reason
1. *PETU* is a rectangle.	1. Given
2. $\overline{UE} \cong \overline{TP}$	2. Diagonals of a rectangle are congruent.
3. $\overline{UE} \cong \overline{CI}$	3. Given
4. $\overline{TP} \cong \overline{CI}$	4. Transitive
5. $\overline{TC} \cong \overline{PI}$	5. Given
6. *PICT* is a parallelogram.	6. If both pairs of opposite sides are congruent, the quadrilateral is a parallelogram.
7. $\angle PET$ is a right angle.	7. A rectangle contains four right angles
8. $\overline{PC} \perp \overline{TI}$	8. Perpendiculars form right angles.
9. *PICT* is a rhombus.	9. If the diagonals of a parallelogram are perpendicular, the parallelogram is a rhombus.

Figure 9.54

Lesson 9-5

1. a. $180(5-2)=540°$
    b. $180(8-2)=1,080°$
    c. $180(17-2)=2,700°$

2. a. $\dfrac{180(6-2)}{6}=\dfrac{720}{6}=120°$

    b. $\dfrac{180(7-2)}{7}=\dfrac{900}{7}=128\dfrac{4}{7}°$

3. $\dfrac{360}{10}=36°$

Circles

Lesson 10-1: Circles, Radii, Chords, and Diameters

A **circle** is defined as the set of all points that are at a fixed distance, called the radius, from a fixed point, called the center. Anyone who has scribed a circle with a compass has used this definition. The point of the compass is held steady at the center of the circle, the arms of the compass are opened to the desired radius, and the pencil arm traces out all the points at that distance, forming the circle.

Figure 10.1

A portion of a circle between two points is called an **arc**. The section of a circle between point A and point B is designated as $\overparen{AB}$. Arcs are measured in degrees, with 360° in the full circle. An arc that measures 180° is a semicircle. Arcs that measure less than 180° are minor arcs; those that measure more the 180° are major arcs. If there is any possibility of confusion about what arc you want to indicate, as there would be here if you wrote $\overparen{CT}$, use another point on the circle. $\overparen{CAT}$ names the arc that goes clockwise from C, through A, to T, and $\overparen{CT}$ will designate the arc from C counterclockwise to T. Generally, the extra point will be added to the major arc.

Figure 10.2

In the coordinate plane, you can develop an equation that describes a circle. Imagine you want a circle with its center at the origin and a radius of five units. Every point on that circle must be five units from the origin. You know that the distance formula is $d = \sqrt{(x_2 - x_1)^2 + (y_2 - y_1)^2}$. Because the radius should be 5, and the origin is the point $(0, 0)$, the equation becomes very simple. Any point (x, y) on the circle must fit the equation $5 = \sqrt{(x-0)^2 + (y-0)^2}$

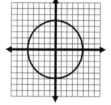

or $5 = \sqrt{x^2 + y^2}$. Squaring both sides eliminates the radical and gives you $x^2 + y^2 = 25$. Changing the radius of the circle will only change the value that was substituted for d, so for any radius r, the equation of a circle of radius r centered at the origin is $x^2 + y^2 = r^2$.

Figure 10.3

Example 1

Find the equation of a circle with radius 7 centered at the origin.

Solution: The basic form of the equation is $x^2 + y^2 = r^2$ and $r = 7$, so the equation is $x^2 + y^2 = 49$.

Example 2

Graph $x^2 + y^2 = 4$.

Solution: $x^2 + y^2 = 4$ is the equation of a circle with its center at the origin, and $r^2 = 4$, so the radius is 2.

Figure 10.4

If the center of the circle is at some point other than the origin—for example, the point $(3, 4)$—and the radius is 5, then $d = \sqrt{(x_2 - x_1)^2 + (y_2 - y_1)^2}$

becomes $5 = \sqrt{(x-3)^2 + (y-4)^2}$. Continue by squaring both sides to get $(x - 3)^2 + (y - 4)^2 = 25$. In general, if the center of the circle is (h, k) and the radius is r, the equation of the circle is $(x - h)^2 + (y - k)^2 = r^2$.

Example 3

Find the equation of a circle of radius 6 with center at $(5, -2)$.

Solution: The basic form of the equation is $(x - h)^2 + (y - k)^2 = r^2$. In this case, $h = 5$, $k = -2$, and $r = 6$, so the equation is $(x - 5)^2 + (y - (-2))^2 = 6^2$ or $(x - 5)^2 + (y + 2)^2 = 36$.

A **chord** is a line segment that has both of its endpoints on the circle. In any circle, many different chords can be drawn, and the length of the chord varies depending upon how far from the center of the circle the chord is located. When you measure the distance of a chord from the center of the circle, you always measure the perpendicular distance. The farther from the center of the circle the chord is located, the shorter the chord will be. Longer chords are closer to the center. The longest possible chord in any circle is the chord that passes right through the center, called the diameter. The diameter is twice as long as the radius.

In circle O, $\overline{AB}$ is a chord and the distance of $\overline{AB}$ from the center is represented by OM. If radius $\overline{OA}$ is drawn, $\triangle OMA$ is a right triangle. The Pythagorean Theorem connects the radius of the circle, OA, the distance of chord from the center, OM, and the length of the chord, because AM is half of AB.

Figure 10.5

Example 4

In a circle of radius 5 cm, a chord 6 cm long is drawn. How far from the center is the chord?

Solution: Using Figure 10.5 as an illustration, $OA = 5$ and $AB = 6$, so $AM = 3$. To find the distance of the chord from the center, OM, use the Pythagorean Theorem.

$$a^2 + b^2 = c^2$$
$$OM^2 + AM^2 = OA^2$$
$$OM^2 + 3^2 = 5^2$$
$$OM^2 + 9 = 25$$
$$OM^2 = 16$$
$$OM = 4$$

The line segments connected to the circle are the radius, the diameter, and other chords. The **radius** is a line segment with one endpoint at the center and one on the circle. Chords, including the diameter, have both endpoints on the circle.

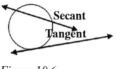

Figure 10.6

The lines associated with the circle are the tangent and the secant. A **tangent** is a line that just touches the circle at exactly one point. The single point at which the tangent meets the circle is called the point of tangency. A **secant** is a line that intersects the circle in two points, or a line that contains a chord.

Example 5

In Figure 10.7, identify:

a. a secant

b. a radius

c. a chord

d. a tangent

e. a diameter

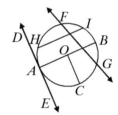

Figure 10.7

Solution:

a. $\overleftrightarrow{FG}$ is a line that intersects the circle at F and at G, so it is a secant. It contains chord $\overline{FG}$.

b. $\overline{OA}$, $\overline{OB}$, and $\overline{OC}$ are all radii. Each segment connects center O with a point on the circle.

c. $\overline{HI}$, $\overline{FG}$, and $\overline{AB}$ are all chords. Their endpoints are points on the circle.

d. $\overleftrightarrow{DE}$ is a line that touches the circle only at point A, so it is a tangent.

e. $\overline{AB}$ is a chord that passes through center O, so it is a diameter.

Lesson 10-1 Review

1. Find the equation of a circle of radius 9, centered at the origin.

2. Graph $x^2 + y^2 = 9$.

3. Find the equation of a circle of radius 12 centered at the point $(-4, 7)$.

4. A chord 24 inches long is drawn 5 inches from the center of circle O. What is the radius of the circle?

5. In the diagram, identify:

    a. $\overline{OQ}$

    b. $\overline{RS}$

    c. $\overleftrightarrow{RS}$

    d. $\overline{PQ}$

    e. $\overline{TV}$

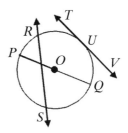

Figure 10.8

Lesson 10-2: Angles in Circles

Drawing radii, diameters, chords, secants, and tangents can create many different types of angles, some of which have specific names. An angle formed by two radii is called a central angle. An angle formed by two chords that has its vertex on the circle is an inscribed angle. Other angles are simply referred to by the lines or segments that form them, for example, an angle formed by a tangent and a secant.

All of these angles have measurements that are related in some way to their intercepted arcs. The intercepted arc is the section of the circle cut off between the sides of the angle. Rather than learn many different rules for all the different angles that can be drawn, organize the angles into four categories, based on where the vertex of the angle is located: at the center, on the circle, inside the circle, and outside the circle.

Inscribed angle

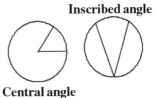

Central angle

Figure 10.9

Central angles, which have their vertex at the center of the circle, have a measurement equal to the measure of the intercepted arc.

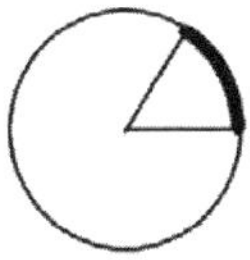

Figure 10.10

Example 1

In circle O, $\overset{\frown}{DT}$ measures $52°$. Find the measure of $\angle DOT$.

Solution: $\angle DOT$ is a central angle. Because its vertex is at the center, $m\angle DOT = m \overset{\frown}{DT} = 52°$.

Example 2

In circle O, $\overline{OH} \perp \overline{OT}$. Find the measure of $\overset{\frown}{HT}$.

Solution: $\overline{OH} \perp \overline{OT}$, so $\angle HOT$ is a right angle. $\angle HOT$ is a central angle, so m$\angle HOT$ = m$\overset{\frown}{HT}$ = 90°.

Inscribed angles are formed by two chords meeting at a point on the circle, and are measured in the same way as angles formed by a radius or diameter meeting a tangent, or by a chord meeting a tangent. All angles with a vertex on the circle are equal to half the measure of the intercepted arc.

Figure 10.11

You can prove that the measure of an inscribed angle is equal to half the measure of its intercepted arc if you consider three cases. If one of the sides of the inscribed angle is a diameter, as $\overline{BOC}$ in Figure 10.12, draw a radius from center O to the end of the other chord, A. $\angle AOC$ is an exterior angle of $\triangle AOB$, so m$\angle AOC$ = m$\angle OAB$ + m$\angle B$. Radii $\overline{OA}$ and $\overline{OB}$ are congruent, so $\triangle AOB$ is isosceles and $\angle OAB \cong \angle B$. There-fore, m$\angle AOC$ = 2m$\angle B$ or m$\angle B$ = $\frac{1}{2}$m$\angle AOC$ = $\frac{1}{2}$m$\overset{\frown}{AC}$.

Figure 10.12

If the sides of the inscribed angle fall on opposite sides of the center, drawing a diameter divides the angle into two angles that fit Case 1. Then:

Figure 10.13

$$\text{m}\angle ABC = \text{m}\angle ABD + \text{m}\angle DBC = \tfrac{1}{2}\text{m}\overset{\frown}{AD} + \tfrac{1}{2}\text{m}\overset{\frown}{DC} = \tfrac{1}{2}\text{m}\overset{\frown}{AC}.$$

If the sides of the inscribed angle fall on the same side of the center, as in Figure 10.14, then: m$\angle ABC$ = m$\angle ABD$ − m$\angle DBC$, and both $\angle ABD$ and $\angle DBC$ fit Case 1. So m$\angle ABC$ = $\frac{1}{2}$m$\overset{\frown}{AD}$ − $\frac{1}{2}$m$\overset{\frown}{DC}$ = $\frac{1}{2}$m$\overset{\frown}{AC}$.

Figure 10.14

When a tangent meets a chord at a point on the circle, two angles are formed. These angles are a linear pair. The obtuse angle, $\angle 2$, is half the measure of the intercepted major arc, and the acute angle, $\angle 1$, is half the measure of the smaller, minor arc.

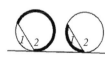

Figure 10.15

Example 3

∠*SUN* is inscribed in a circle. If $\overset{\frown}{SN}$ measures 84°, find the measure of ∠*SUN*.

Solution: The inscribed angle, ∠*SUN*, has its vertex, *U*, on circle O. The measure of the angle is half the measure of the intercepted arc, $\overset{\frown}{SN}$,

so m∠*SUN* = $\frac{1}{2}$ m $\overset{\frown}{SN}$ = $\frac{1}{2}$(84°) = 42°.

Figure 10.16

Example 4

$\overleftrightarrow{RD}$ is tangent to a circle at point *A*, and $\overline{IA}$ is a chord. If m $\overset{\frown}{AI}$ = 2*x* – 20 and m $\overset{\frown}{IOA}$ = 5*x* + 30, find the measure of ∠*RAI*.

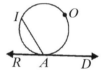

Figure 10.17

Solution: $\overset{\frown}{AI}$ and $\overset{\frown}{IOA}$ together make up the entire circle, and therefore add to 360°.

$$2x - 20 + 5x + 30 = 360$$
$$7x + 10 = 360$$
$$7x = 350$$
$$x = 50$$

m $\overset{\frown}{AI}$ = 2*x* – 20 = 2(50) – 20 = 80°. The measure of ∠*RAI* is half the measure of $\overset{\frown}{AI}$, so m∠*RAI* = $\frac{1}{2}$(80) = 40°.

Example 5

Show that a tangent and a radius (or a diameter) drawn to the point of tangency are always perpendicular.

Figure 10.18

Solution: Diameter $\overline{MT}$ passes through center A and intersects tangent $\overleftrightarrow{NL}$ at *T*, the point of tangency. ∠*MTL* has its vertex on the circle and is equal to half the measure of $\overset{\frown}{MET}$. Because $\overline{MT}$ is a diameter, $\overset{\frown}{MET}$ is a semicircle. Therefore $\angle MTL = \frac{1}{2}(180) = 90°$. $\overrightarrow{MT}$ and $\overrightarrow{NL}$ form a right angle, so $\overrightarrow{MT} \perp \overrightarrow{NL}$.

When two chords or secants intersect within the circle, they form four angles. There are two pairs of vertical angles, and vertical angles are congruent. Each angle in the pair intercepts an arc, and the measure of the either angle in the pair is the average of those two intercepted arcs.

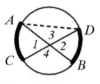

Figure 10.19

Suppose chords $\overline{AB}$ and $\overline{CD}$ intersect inside the circle at P. If you draw $\overline{AD}$, $m\angle ADC = \frac{1}{2}m\overset{\frown}{AC}$ and $m\angle BAD = \frac{1}{2}m\overset{\frown}{BD}$. Because $\angle 1$ is an exterior angle of

$\triangle APD$, $m\angle 1 = m\angle ADC + m\angle BAD = \frac{1}{2}m\overset{\frown}{AC} + \frac{1}{2}m\overset{\frown}{BD} = \frac{1}{2}(m\overset{\frown}{AC} + m\overset{\frown}{BD})$. An angle with its vertex inside the circle is equal to one-half the sum of the intercepted arcs.

Example 6

Chords $\overline{MS}$ and $\overline{AI}$ intersect at point R in the interior of the circle. $m\angle 1 = 2x + 4$, $m\overset{\frown}{MA} = 3x - 1$ and $m\overset{\frown}{IS} = 26°$. Find the measure of $\angle ARS$.

Figure 10.20

Solution: The measure of $\angle 1$ is one-half the sum of $m\overset{\frown}{MA}$ and $m\overset{\frown}{IS}$. Therefore,

$2x + 4 = \frac{1}{2}(3x - 1 + 26)$
$2x + 4 = \frac{1}{2}(3x + 25)$
$4x + 8 = 3x + 25$
$\qquad x = 17$

$m\angle 1 = 2x + 4 = 2(17) + 4 = 38°$. $\angle ARS$ is supplementary to $\angle 1$, so $m\angle ARS = 180 - 38 = 142°$.

When two tangents, two secants, or a tangent and a secant are drawn from a single point, an angle is formed that has its vertex outside the circle. These angles each intercept two arcs of the circle, and the measure of the angle is one-half the difference of the arcs.

Figure 10.21

For an angle formed by two secants, draw a chord as shown, and $m\angle 1 = m\angle P + m\angle 2$, so $m\angle P = m\angle 1 - m\angle 2$. Both $\angle 1$ and $\angle 2$ are inscribed angles, measured by half their intercepted arcs, so $m\angle P$ is half the difference of the arcs.

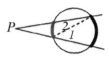

Figure 10.22

The proof for an angle formed by a tangent and a secant is similar except that $\angle 2$ is an inscribed angle and $\angle 1$ is an angle formed by a tangent and a chord.

Figure 10.23

For an angle formed by two tangents, both $\angle 1$ and $\angle 2$ are formed by tangent and chord. For the angle formed by two tangents, it is true that the two intercepted arcs make up the entire circle, and therefore add to 360°. Remember that the measure of the angle is half of the difference of the angles, not half of the sum, and don't jump to conclusions. The arcs will not generally be equal, even if the figure seems to give that impression.

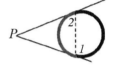

Figure 10.24

Example 7

Secants $\overleftrightarrow{NB}$ and $\overleftrightarrow{NS}$ are drawn to the circle as shown. $m\overarc{OD} = x$, $m\angle N = x + 5$, and $m\overarc{BS} = 58°$. Find the measure of $\overarc{OD}$.

Solution: The measure of $\angle N$ is one-half the difference of the measures of $\overarc{BS}$ and $\overarc{OD}$.

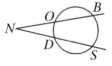

Figure 10.25

$$m\angle N = \tfrac{1}{2}\left(m\overarc{BS} - m\overarc{OD}\right)$$
$$x + 5 = \tfrac{1}{2}(58 - x)$$
$$2x + 10 = 58 - x$$
$$3x = 48$$
$$x = 16$$

The measure of $\overarc{OD}$ is 16°.

Example 8

Secant $\overleftrightarrow{YM}$ and tangent $\overleftrightarrow{YS}$ are drawn to the circle as shown. $m\overparen{MA} = 34°$ and $m\angle Y = 77°$. Find $m\overparen{SM}$.

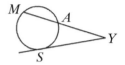

Figure 10.26

Solution: At first it may appear that there is not enough information to solve the problem, but remember that

$m\overparen{MA} + m\overparen{SM} + m\overparen{AS} = 360°$, so $34° + m\overparen{SM} + m\overparen{AS} = 360°$ or $m\overparen{SM} + m\overparen{AS} = 326°$. In addition, you know that

$m\angle Y = \frac{1}{2}\left(m\overparen{SM} - m\overparen{AS}\right)$, so $77° = \frac{1}{2}\left(m\overparen{SM} - m\overparen{AS}\right)$, or

$154° = m\overparen{SM} - m\overparen{AS}$. These two equations form a system that can be solved to find the measure of both arcs.

$$m\overparen{SM} + m\overparen{AS} = 326°$$
$$m\overparen{SM} - m\overparen{AS} = 154°$$
$$2m\overparen{SM} \quad\quad = 480°$$
$$m\overparen{SM} \quad\quad = 240°$$
$$240° + m\overparen{AS} = 326°$$
$$m\overparen{AS} = 86°$$

Example 9

Tangents $\overleftrightarrow{ET}$ and $\overleftrightarrow{EJ}$ are drawn to the circle as shown. $m\angle E = 30°$. Find the measure of $\overparen{JRT}$.

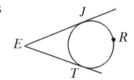

Figure 10.27

Solution: $m\overparen{JT} + m\overparen{JRT} = 360°$, or

$m\overparen{JT} = 360° - m\overparen{JRT} \cdot m\angle E = \frac{1}{2}\left(m\overparen{JRT} - m\overparen{JT}\right)$.

$m\angle E = 30°$, so this becomes $30° = \frac{1}{2}\left(m\overparen{JRT} - m\overparen{JT}\right)$ or

$60° = m\overparen{JRT} - m\overparen{JT}$.

Substitute to get $60° = m\widehat{JRT} - \left(360° - m\widehat{JRT}\right)$ or

$60° = m\widehat{JRT} - 360° + m\widehat{JRT} = 2m\widehat{JRT} - 360°$. Solving, you get

$420° = 2m\widehat{JRT} \Rightarrow m\widehat{JRT} = 210°$.

Lesson 10-2 Review

1. In circle O, $\angle DOT$ measures 89°. Find the measure of $\widehat{DT}$.

2. $\angle SUN$ is inscribed in a circle. If $\angle SUN$ measures 135°, find the measure of $\widehat{SN}$.

3. $\overleftrightarrow{RD}$ is tangent to a circle at point A, and $\overline{IA}$ is a chord. If $m\angle RAI = 10x - 13$ and $m\angle IAD = 3x - 2$, find the measure of minor arc $\widehat{AI}$.

4. Chords $\overline{MS}$ and $\overline{AI}$ intersect at point R in the interior of the circle. $m\angle MRA = 7x + 1$, and $m\angle SRI = 9x - 15$. If $m\widehat{MA} = 80°$, find $m\widehat{IS}$.

Use Figure 10.28 for questions 5, 6, and 7.

5. Secants $\overleftrightarrow{NA}$ and $\overleftrightarrow{NK}$ are drawn to the circle as shown. $\widehat{AK}$ is three times as large as $\widehat{CS}$, and $m\angle ANK = 54°$. Find the measure of $\widehat{CS}$.

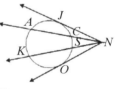

Figure 10.28

6. Secant $\overleftrightarrow{NK}$ and tangent $\overleftrightarrow{NJ}$ are drawn to the circle as shown. $m\widehat{SOK} = 160°$, and $m\angle KNJ = 80°$. Find $m\widehat{KAJ}$.

7. Tangents $\overleftrightarrow{NJ}$ and $\overleftrightarrow{NO}$ are drawn to the circle as shown. $m\angle JNO = 150°$. Find the measure of $\widehat{JCO}$.

Lesson 10-3: Segment Relationships

In most of the relationships described in this lesson, two lines—tangents or secants—will be drawn to the circle through a single point outside the circle. The external segments of interest are the segments that have that external point as one endpoint and a point of the circle as the other endpoint. The internal segment is the chord the secant contains, with its endpoints on the circle. Tangents, except for the point of tangency that they share with the circle, are outside the circle.

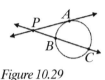

Figure 10.29

In Figure 10.29, the external segment of tangent $\overrightarrow{PA}$ is the segment $\overline{PA}$, from the external point P to the point A on the circle. The external segment of secant $\overrightarrow{PC}$ is $\overline{PB}$, and the internal segment is chord $\overline{BC}$.

When two tangents are drawn to a circle from the same point, you can show that the external segments are congruent. Draw the radii $\overline{OA}$ and $\overline{OB}$ to the two points of tangency, and draw $\overline{PO}$. $\overline{OA} \perp \overline{PA}$ and

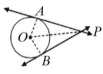

Figure 10.30

$\overline{OB} \perp \overline{PB}$, because the radius to the point of tangency is perpendicular to the tangent, and so $\triangle PAO$ and $\triangle PBO$ are right triangles (◣). $\overline{PO} \cong \overline{PO}$ (H) and, because all radii of a circle are congruent, $\overline{OA} \cong \overline{OB}$ (L).

The two right triangles are therefore congruent by HL, and, because $\triangle PAO \cong \triangle PBO$, you can conclude that $\overline{PA} \cong \overline{PB}$ by CPCTC.

Example 1

Tangents $\overrightarrow{PA}$ and $\overrightarrow{PB}$ are drawn to circle O and chord $\overline{AB}$ connects the points of tangency. Prove that $\angle PAB \cong \angle PBA$.

Figure 10.31

Solution:

Statement	Reason
1. Tangents $\overrightarrow{PA}$ and $\overrightarrow{PB}$ are drawn to circle O.	1. Given
2. $\overline{PA} \cong \overline{PB}$	2. If two tangents are drawn to a circle from the same point, the external segments are congruent.
3. $\angle PAB \cong \angle PBA$	3. If two sides of a triangle ($\triangle PAB$) are congruent, the angles opposite those sides are congruent.

Figure 10.32

In Figure 10.33, secants $\overrightarrow{PA}$ and $\overrightarrow{PD}$ are drawn.

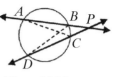

$\overline{PB}$ and $\overline{PC}$ are the external segments, $\overline{BA}$ and $\overline{CD}$ are the internal segments, and $\overline{PA}$ and $\overline{PD}$, the combination of the internal and external segments, are the secant segments.

Figure 10.33

Draw $\overline{AC}$ and $\overline{BD}$ to form $\triangle PAC$ and $\triangle PDB$. $\angle APD$ is in both triangles, and $\angle APD \cong \angle APD$ (A). $\angle PAC$ and $\angle PDB$ are inscribed angles and intercept the same arc, so m$\angle PAC$ = m$\angle PDB$ (A). That's enough to prove that $\triangle PAC \sim \triangle PDB$ by AA, so you can conclude that corresponding sides are in proportion. $\frac{PA}{PD} = \frac{PC}{PB}$ means that the lengths of the secant segments are proportional to the lengths of their external segments.

Example 2

Secants $\overrightarrow{PA}$ and $\overrightarrow{PD}$ are drawn to a circle, as shown. If $AB = x - 2$, $BP = 5$, $DC = 4$, and $CP = x - 3$, find the length of $\overline{AP}$.

Solution: Because the two secants are drawn to the circle from the same point, P, you

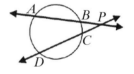

know that $\dfrac{AP}{DP} = \dfrac{CP}{BP}$. AP and DP are not directly given, but can be expressed from the given information. $AP = AB + BP = x - 2 + 5 = x + 3$. $DP = DC + CP = 4 + x - 3 = 1 + x$. Cross-multiplying and substituting, you get:

$AP \cdot BP = DP \cdot CP$

$(x + 3)(5) = (1 + x)(x - 3)$

$5x + 15 = x^2 - 2x - 3$

$x^2 - 7x - 18 = 0$

The quadratic equation can be factored to $(x + 2)(x - 9) = 0$, and setting each factor equal to zero produces two solutions: $x + 2 = 0 \Rightarrow x = -2$ and $x - 9 = 0 \Rightarrow x = 9$. A negative value does not make sense as a length, so accept the solution $x = 9$. The length of $\overline{AP}$ is $AP = AB + BP = x - 2 + 5 = x + 3 = 9 + 3 = 12$.

When a secant and a tangent are drawn from the same point, it is possible to find similar triangles, if you draw chords $\overline{AB}$ and $\overline{AC}$. The goal is to prove that $\triangle PAC \sim \triangle PBA$. Once again $\angle P$ is in both triangles, and $\angle P \cong \angle P$ (A). $\angle PAC$ is an angle formed by a tangent and a chord, with its vertex on the circle, so its measure is one-half the measure of its intercepted arc, $\overset{\frown}{AC}$. $\angle PBA$ is an inscribed angle, equal to one-half the measure of $\overset{\frown}{AC}$, so $\angle PAC \cong \angle PBA$ (A). $\triangle PAC \sim \triangle PBA$ by AA, and so corresponding sides are in proportion. Using just the sides that were part of the original figure, you get $\frac{PC}{PA} = \frac{PA}{PB}$. The tangent segment is the mean proportional between the secant segment and the external segment of the secant.

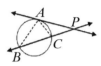

Figure 10.35

Example 3

Tangent $\overrightarrow{PA}$ and secant $\overrightarrow{PB}$ are drawn to a circle as shown. If $CB = 3x + 1$, $PC = 3x - 1$, and $AP = 12$, find the length of $\overline{PB}$.

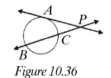

Figure 10.36

Solution: AP is the mean proportional between PB and PC, so $\frac{PC}{PA} = \frac{PA}{PB}$. $PB = PC + CB = 3x - 1 + 3x + 1 = 6x$,

so the proportion becomes $\frac{6x}{12} = \frac{12}{3x - 1}$. Cross-multiplying gives $6x(3x - 1) = 12^2$ or $18x^2 - 6x - 144 = 0$. The terms have a common factor of 6, which can be factored out for $6(3x^2 - x - 24) = 0$ and both sides divided by 6 for $3x^2 - x - 24 = 0$. Further factoring gives you $(3x + 8)(x - 3) = 0$ and solving produces $3x + 8 = 0 \Rightarrow x = \dfrac{-8}{3}$, which does not make sense as a length, or $x - 3 = 0 \Rightarrow x = 3$. If $x = 3$, $PB = PC + CB = 3x - 1 + 3x + 1 = 6x = 6(3) = 18$.

When two chords intersect within a circle, each chord divides the other into two segments, but be aware that *divide* does not mean *bisect*. It is possible for one chord to bisect another, but the only time both chords are bisected is when both chords are diameters. For other chords, it's possible to show similar triangles and therefore segments that are in proportion.

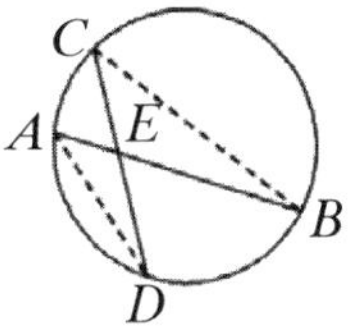

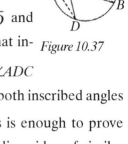

If chords $\overline{AB}$ and $\overline{CD}$ intersect at E, draw $\overline{AD}$ and $\overline{BC}$. $\angle ADC$ and $\angle ABC$ are both inscribed angles that in-

Figure 10.37

tercept $\overset{\frown}{AC}$, so both measure $\frac{1}{2}m\overset{\frown}{AC}$. That means $\angle ADC$ $\cong \angle ABC$ (A). By similar logic, $\angle DAB$ and $\angle DCB$ are both inscribed angles equal to $\frac{1}{2}m\overset{\frown}{DB}$, and so $\angle DAB \cong \angle DCB$ (A). This is enough to prove that $\triangle ADE \sim \triangle CBE$, and then, because corresponding sides of similar triangles are in proportion, $\frac{AE}{CE} = \frac{DE}{BE}$ or $AE \cdot BE = CE \cdot DE$. This cross-multiplied version tells you that the product of the lengths of the two segments of the chord is always the same.

Example 4

Chords $\overline{AB}$ and $\overline{CD}$ intersect at E. $AE = 3$, $DE = x - 3$, $CE = x + 2$, and $BE = x + 5$. Find the length of $\overline{CD}$.

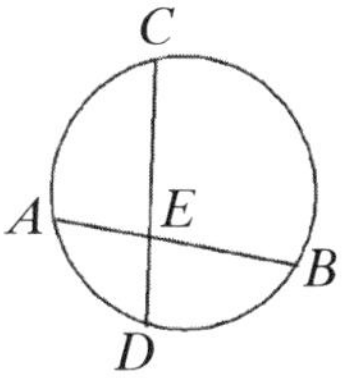

Solution: When two chords intersect, the product of the lengths of the two segments of each chord is constant, so $AE \cdot BE = CE \cdot DE$. Substituting the given information, you have $3(x + 5) = (x - 3)(x + 2)$ or $3x + 15 = x^2 - x - 6$. Collecting terms on one side gives you $x^2 - 4x - 21 = 0$, which can be factored as $(x + 3)(x - 7) = 0$. Setting each factor equal to zero, you get two solutions: $x + 3 = \Rightarrow x = -3$ and $x - 7 = 0 \Rightarrow x = 7$. Only the positive solution will make sense, so accept $x = 7$, and calculate the length of $\overline{CD}$:
$CD = CE + DE = x + 2 + x - 3 = 2x - 1 = 2(7) - 1 = 13$.

Figure 10.38

Lesson 10-3 Review

1. Tangents $\overrightarrow{PC}$ and $\overrightarrow{PD}$ are drawn to circle O as shown in Figure 10.39.
   If $m\angle CAB = 3(x + 1)$ and $m\angle DBA = 143 - x$, find the measure of $\angle P$.

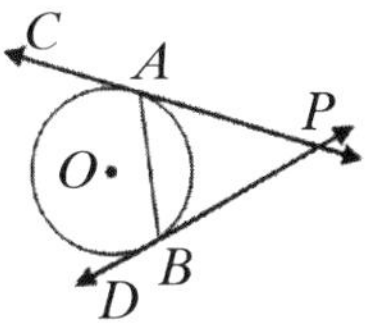

Figure 10.39

2. Secants $\overleftrightarrow{PB}$ and $\overleftrightarrow{PD}$ are drawn to a circle, as shown in Figure 10.40. If $BP = 3x$, $CP = x$, $DP = 4x$, and $EP = 3$, find the length of $\overline{BC}$.

3. Tangent $\overrightarrow{PA}$ and secant $\overrightarrow{PB}$ are drawn to a circle as shown in Figure 10.40. If $PB = x + 2$, $PC = x - 3$, and $PA = x - 1$, find the length of $\overline{BC}$.

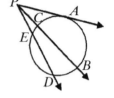

Figure 10.40

4. Chords $\overline{AB}$ and $\overline{CD}$ intersect at E. $AE = 6$, $DE = x + 2$, $CE = 9$, and $BE = 2x$. Find the length of $\overline{AB}$.

5. Prove that a diameter perpendicular to a chord bisects the chord. (Hint: Draw radii to the endpoints of the chord.)

Answer Key

Lesson 10-1

1. $x^2 + y^2 = 81$
2. Center $(0,0)$, radius $= 3$

Figure 10.41

3. $(x+4)^2 + (y-7)^2 = 144$
4. $a^2 + b^2 = c^2$
   $12^2 + 5^2 = c^2$
   $144 + 25 = 169 = c^2$
   $c = 13$
   The radius of the circle is 13.
5. a. radius   b. chord   c. secant   d. diameter   e. tangent

Lesson 10-2

1. $m \overarc{DT} = 89°$

2. $m \overarc{SN} = 270°$

3. $m\angle RAI + m\angle IAD = 180$

   $10x - 13 + 3x - 2 = 180$

   $13x - 15 = 180$

   $13x = 195$

   $x = 15$

   $m\angle RAI = 10x - 13 = 10(15) - 13 = 137.$

   $m\overarc{AI} = 2m\angle RAI = 2(137) = 274°.$

4. $7x + 1 = 9x - 15$

   $16 = 2x$

   $x = 8$

   The two angles each measure $7(8) + 1 = 9(8) - 15 = 57°$, and either of the angles is equal to one-half the sum of the intercepted arcs.

   $m\angle SRI = \frac{1}{2}\left(m\overarc{MA} + m\overarc{IS}\right)$

   $57° = \frac{1}{2}\left(80° + y\right)$

   $114 = 80 + y$

   $y = 34°$

   $m\overarc{IS} = 34°$

5. $m\angle ANK = \frac{1}{2}\left(m\overarc{AK} - m\overarc{CS}\right)$

   $54° = \frac{1}{2}(3x - x)$

   $108 = 2x$

   $x = 54$

   $m\overarc{CS} = 54°$

6. $m\angle KNJ = \frac{1}{2}\left(m\overarc{KAJ} - m\overarc{JCS}\right)$ $m\overarc{SOK} + m\overarc{KAJ} + m\overarc{JCS} = 360°$

   $80° = \frac{1}{2}(x - y)$ $160 + x + y = 360$

   $160 = x - y$ $x + y = 200$

   $$x - y = 160$$
   $$x + y = 200$$
   $$2x = 360$$
   $$x = 180$$
   $$y = 20$$
   $$m\overarc{KAJ} = 180°$$

7. $m\angle JNO = \frac{1}{2}\left(m\widehat{JAO} - m\widehat{JCO}\right)$ $\qquad$ $m\widehat{JAO} + m\widehat{JCO} = 360°$

$\quad 150° = \frac{1}{2}(x - y)$ $\qquad\qquad\qquad\qquad$ $x + y = 360$

$\quad 300 = x - y$

$$x - y = 300$$
$$x + y = 360$$
$$2x = 660$$
$$x = 330$$
$$y = 30$$
$$m\widehat{JCO} = 30°$$

Lesson 10-3

1. $\overline{PC} \cong \overline{PD}$ so $\angle PCD \cong \angle PDC$, and therefore $\angle CAB \cong \angle DBA$.

$$3(x + 1) = 143 - x$$
$$3x + 3 = 143 - x$$
$$4x = 140$$
$$x = 35$$
$$m\angle CAB = m\angle DBA = 108°$$
$$m\angle PAB = m\angle PBA = 72°$$
$$m\angle P = 180 - 2(72) = 36°$$

2. $BP \cdot CP = DP \cdot EP$

$$3x \cdot x = 4x \cdot 3$$
$$3x^2 = 12x$$
$$3x^2 - 12x = 0$$
$$3x(x - 4) = 0$$
$$3x = 0 \Rightarrow x = 0$$
$$x - 4 = 0 \Rightarrow x = 4$$
$$BC = BP + CP = 3(4) + 4 = 16$$

3.
$$PA^2 = PB \cdot PC$$
$$(x-1)^2 = (x+2)(x-3)$$
$$x^2 - 2x + 1 = x^2 - x - 6$$
$$-2x + 1 = -x - 6$$
$$7 = x$$
$$BC = PB + PC = (7+2) + (7-3) = 9 + 4 = 13$$

4. $AE \cdot BE = CE \cdot DE$
$$6 \cdot 2x = 9(x+2)$$
$$12x = 9x + 18$$
$$3x = 18$$
$$x = 6$$
$$AB = AE + BE = 6 + 2(6) = 18$$

5.

Statement	Reason
1. Circle O with diameter $\overline{CD}$ perpendiculat to chord $\overline{AB}$ at E.	1. Given
2. $\angle OEA$ and $\angle OEB$ are right angles.	2. Perpendiculars form right angles.
3. $\angle OEA \cong \angle OEB$	3. All right angles are congruent.
4. Draw radii $\overline{OA}$ and $\overline{OB}$	4. Two points determine a unique line.
5. $\triangle OAE$ and $\triangle OBE$ are right triangles. (◣)	5. A triangle that contains one right angle is a right triangle.
6. $\overline{OA} \cong \overline{OB}$ (H)	6. All radii of the circle are congruent.
7. $\overline{OE} \cong \overline{OE}$ (L)	7. Reflexive
8. $\triangle OAE \cong \triangle OBE$	8. HL
9. $\overline{AE} \cong \overline{EB}$	9. CPCTC
10. $\overline{CD}$ bisects $\overline{AB}$	10. A bisector divides the segment into two congruent segments.

Figure 10.42

Area

Lesson 11-1: Quadrilaterals

Perimeter is a linear measurement, an adding up of the lengths of the line segments that form the sides of a polygon. Area, by contrast, measures a surface, a plane. Whereas perimeter is measured in linear units, such as centimeters or inches, area is measured in square units. A square unit can be imagined as a tile measuring one unit on each side. The area of the polygon is the number of such tiles required to cover interior of the polygon.

For rectangles and squares, such a tiling is simple. If a rectangle measures 3 cm by 5 cm, it can be covered by three rows of five tiles, representing an area of 15 square centimeters. If the dimensions of the rectangle are not integers, the count is more complicated because it involves fractions of tiles, but the formula for the area of a rectangle is easy to apply, whatever the dimensions.

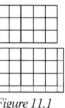

Figure 11.1

The **area of a rectangle** is equal to the product of the length and the width. The length and width are sometimes designated as the base and the height, so the formula for the area of a rectangle may be given as $A = lw$ or $A = bh$. The length and width of a square are identical measurements, so the formula for the area of a square is often given as $A = s^2$.

Example 1

The length and width of a rectangle are consecutive integers. If the area of the rectangle is 72 square inches, find the dimensions of the rectangle.

Solution: Let the dimensions of the rectangle be represented by x and $x + 1$. Then $A = lw$ becomes $72 = x(x + 1)$. Simplify and move all non-zero terms to one side to get the quadratic equation $x^2 + x - 72 = 0$. Solving by factoring, $(x + 9)(x - 8) = 0$ means $x = -9$ or $x = 8$. A negative solution is not possible, so $x = 8$ inches is one dimension of the rectangle. The other dimension is $x + 1 = 9$ inches.

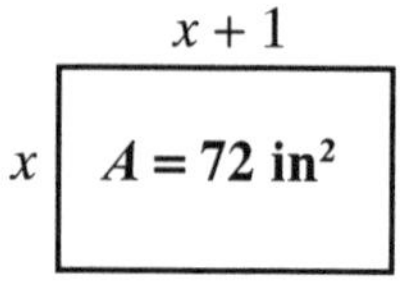

$x + 1$

x | $A = 72$ in²

Figure 11.2

Example 2

The diagonal of a square is 24 cm. Find the area of the square.

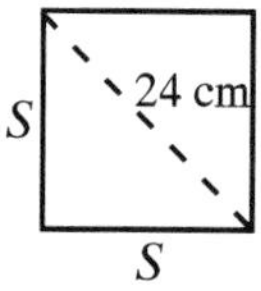

S | 24 cm

S

Figure 11.3

Solution: Given the length of the diagonal, you can find the length of a side by using the Pythagorean Theorem, or by 45–45–90 right triangle relationships. By either method, you find that $2s^2 = 24^2$ or $s = 12\sqrt{2}$. The area of the square, $A = s^2$, becomes

$$A = \left(12\sqrt{2}\right)^2 = 12^2 \cdot 2 = 288 \text{ cm}^2 \cdot$$

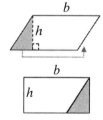

Figure 11.4

For parallelograms, as for most other polygons, the rule for calculating area is derived from known formulas. In the parallelogram, if a perpendicular is drawn from one side to the opposite, parallel side, the length of the side is the base and the length of the perpendicular segment is the height. If the section to one side of the perpendicular is moved to the other end of the parallelogram, a rectangle is formed. The area of the rectangle is equal to the area of the parallelogram, so the **area of the parallelogram** can be given as $A = bh$, where b is the length of a side and h is the length of a perpendicular to that side from its opposite side.

Example 3

The area of a parallelogram is 2,646 cm². If $b = x + 13$ and $h = x - 8$, find the length of the base.

Solution: $A = bh$, so $2{,}646 = (x + 13)(x - 8) = x^2 + 5x - 104$. Collect terms to produce the quadratic equation $x^2 + 5x - 2{,}750 = 0$, and solve. $(x + 55)(x - 50) = 0$ gives two solutions. $x + 55 = 0 \Rightarrow x = -55$ is not a reasonable solution, so $x - 50 = 0 \Rightarrow x = 50$. Therefore the base is $50 + 13 = 63$ cm.

If the height is not known, the area can still be computed if the lengths of the sides and the measure of the acute angle between them are known. If a and b are the lengths of the sides of the parallelogram, drawing the height from a vertex creates a right triangle, and the trigonomet-

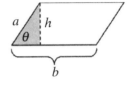

Figure 11.5

ric ratios tell you that $\sin\theta = \frac{\text{opposite}}{\text{hypotenuse}}$. The opposite side is the length of the height, and the hypotenuse is the length of side a, so $\sin\theta = \frac{h}{a}$.

Multiply both sides by a, and $h = a \sin \theta$. Substitute for h in the area formula, and $A = b(a \sin \theta) = ab \sin \theta$. So the area of a parallelogram can be found by multiplying the lengths of the adjacent sides and the sine of the acute angle between them.

Example 4

Parallelogram $\square WXYZ$ has sides $WX = 12$ cm and $XY = 18$ cm, and an area of 108 cm^2. Find the measure of $\angle X$.

Solution: $A = ab \sin \theta$, so $108 = 12 \cdot 18 \sin(\angle X)$ or $108 = 216$ $\sin(\angle X)$. Dividing both sides by 216 tells you that $\sin(\angle X) = \frac{1}{2}$. You may have this value committed to memory, but if not, you can use your calculator to find that $m\angle X = \sin^{-1}\left(\frac{1}{2}\right) = 30°$.

In a rhombus, a parallelogram with four congruent sides, the diagonals, d_1 and d_2, are perpendicular, and therefore divide the rhombus into four congruent right triangles. If those triangles are split apart and reassembled, they create two rectangles, each with length and width of $\frac{1}{2}d_1$ and $\frac{1}{2}d_2$. The total area of those two rectangles will equal the area of the

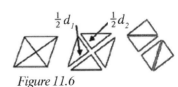

Figure 11.6

rhombus. $A = 2(lw) = 2\left(\frac{1}{2}d_1 \cdot \frac{1}{2}d_2\right) = \frac{1}{2}d_1 d_2$. The **area of a rhombus** is equal to one-half the product of the lengths of the diagonals.

Example 5

Rhombus $\square ABCD$ has sides of length 10 cm and a height of 9.6 cm. If diagonal $\overline{AC}$ is 12 cm long, find the length of $\overline{BD}$.

Solution: Using the parallelogram formula (because every rhombus is a parallelogram), the area of the rhombus is $A = bh = 10(9.6) = 96$ cm^2. But because the area of the rhombus is also $A = \frac{1}{2}d_1 d_2$, you can substitute to get $96 = \frac{1}{2}(12)d_2 = 6d_2$. Dividing by 6 tells you $BD = 16$ cm.

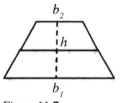

Figure 11.7

To find a formula for the area of a trapezoid, label the bottom base b_1 and the top base b_2. The height is the perpendicular distance between b_1 and b_2. Draw the median of the trapezoid, which is parallel to the bases. The parallel lines—the bases and the median—divide the non-parallel sides in half, so they also divide the height in half.

Cut the trapezoid along the median, and rotate the top section until it sits beside the bottom section. The result is a parallelogram whose base is $b_1 + b_2$ and whose height is half the height of the trapezoid.

Figure 11.8

Substitute into the formula for the area of a parallelogram. $A = bh = (b_1 + b_2)\left(\frac{1}{2}h\right)$. The **area of a trapezoid** is $A = \frac{1}{2}(b_1 + b_2)h$. You can remember this as the average of the bases times the height, or as the length of the median times the height.

Example 6

Trapezoid *ABCD* has a height of 8 inches and an area of 124 square inches. If $\overline{AB} \parallel \overline{CD}$, $AB = 3x - 1$, and $CD = 4x - 3$, find the length of $\overline{AB}$.

Solution: Substitute in $A = \frac{1}{2}(b_1 + b_2)h$ to get

$124 = \frac{1}{2}(3x - 1 + 4x - 3) \cdot 8$. Simplifying gives you $124 = 4(7x - 4)$ or $124 = 28x - 16$. Solve to get $28x = 140$ and $x = 5$. Then $AB = 3(5) - 1 = 14$ inches.

Example 7

Parallelogram $\square WXYZ$ is drawn inside trapezoid *ABCD*, with *W* on $\overline{AB}$ and *Y* on $\overline{CD}$, and $\overline{WY} \perp \overline{AB}$. $AB = 4$ cm, $CD = 12$ cm, $WY = 6$ cm, $WZ = 2$ cm, $WX = 5$ cm, and $m\angle ZWX = 30°$. Find the area of the shaded region.

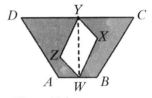

Figure 11.9

Solution: The area of the trapezoid *ABCD* is

$A = \frac{1}{2}(b_1 + b_2)h = \frac{1}{2}(4 + 12) \cdot 6 = 48$ cm^2. The area of the

parallelogram $\square WXYZ$ is $A = ab\sin\theta = 2 \cdot 5 \cdot \sin 30° = 10 \cdot \frac{1}{2} = 5$ cm^2. The area of the shaded region is the area of the trapezoid *ABCD* minus the area of the parallelogram $\square WXYZ$. The shaded area is $48 - 5 = 43$ cm^2

Lesson 11-1 Review

1. The length of a rectangle is $3x$ feet and the width is $3 + x$ feet. If the area of the rectangle is 84 square feet, find the dimensions of the rectangle.

2. Show that, in any square, the area of the square is equal to one-half the square of the diagonal.

3. The area of parallelogram $\square ABCD$ is equal to the area of rhombus $\square WXYZ$, and $AD = 14$. If diagonals $\overline{WY}$ and $\overline{XZ}$ are congruent to $\overline{AB}$ and $\overline{AD}$, respectively, find the length of the altitude to $\overline{AB}$.

4. Rhombus $\square ABCD$ has sides of length 13 cm. If diagonal $\overline{AC}$ is 10 cm long, find the length of $\overline{BD}$ and the area of the rhombus.

5. Trapezoid $ABCD$ has an area of 35 square meters. If the height is equal to the longer base, and the shorter base is 3 meters, find the height.

6. $\square ABCD$ is a rhombus with $AC = 8$ cm and $BD = 14$ cm. Each of the unshaded boxes is a square of side 2 cm. Find the area of the shaded region.

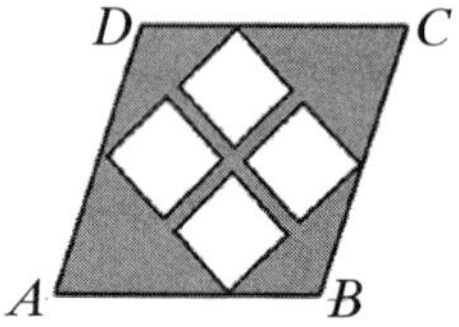

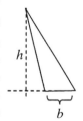

Figure 11.10

Lesson 11-2: Triangles

If a triangle is duplicated, and the copy rotated 180° to place beside the original, a parallelogram is formed with base and height equal to the base and height of the triangle. Because two triangles make up the parallelogram, the area of each triangle is half the area of the parallelogram.

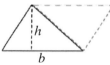

Figure 11.11

The most common formula for the area of a triangle is $A = \frac{1}{2}bh$, where b is the length of any side of the triangle and h is the length of an altitude drawn from the opposite vertex, perpendicular to the side. At times, it may be necessary to drop the perpendicular outside the triangle. You will want to extend the side until it intersects the altitude, so that you can measure the height, but remember that the length of the base is only the length of the original side, not the extension.

Figure 11.12

Example 1

In $\triangle RST$, $RS = 4.2$ cm. The altitude to side $\overline{RS}$ measures 6.8 cm and the altitude to side $\overline{ST}$ measures 8.4 cm. Find the length of $\overline{ST}$.

Solution: Using $\overline{RS}$ and the altitude drawn to $\overline{RS}$,

$A = \frac{1}{2}bh = \frac{1}{2}(4.2)(6.8) = 14.28 \text{ cm}^2$. The area will be the same if

calculated using $\overline{ST}$ and the altitude drawn to $\overline{ST}$:

$14.28 \text{ cm}^2 = \frac{1}{2}(x)(8.4)$ so $14.28 = 4.2x$ and $x = 3.4$ cm.

Example 2

In $\triangle ABC$, altitude $\overline{BD} \perp \overline{AC}$. $BD = 4x - 7, AC = 8x$, and the area
of $\triangle ABC$ is 260 square units, find the length of $\overline{BD}$.

Solution: $A = \frac{1}{2}bh = \frac{1}{2}(8x)(4x-7) = 260$. Simplifying gives you the
quadratic equation $4x(4x - 7) = 260$ or $16x^2 - 28x - 260 = 0$.
Divide through by 4 to make the equation easier to factor.
$4x^2 - 7x - 65 = (4x + 13)(x - 5) = 0$. Setting $4x + 13$ equal to zero
gives a negative solution, which is rejected, but $(x - 5) = 0$ implies
$x = 5$. Then $BD = 4x - 7 = 4(5) - 7 = 13$ units.

The legs of a right triangle are perpendicular, so if one leg is chosen
as the base, the other is the altitude or height. Substituting into the traditional formula, you find that the area of a right triangle is one-half the
product of the lengths of the legs. If the legs are designated as a and b,
$A = \frac{1}{2}ab$.

Example 3

In right triangle $\triangle RST$, with $\overline{RS} \perp \overline{ST}$, $ST = 10$ cm, and RS and
RT are consecutive even integers. Find the area of $\triangle RST$.

Solution: $RS = x$ and $RT = x + 2$.
Apply the Pythagorean Theorem to find the length of RS.
$10^2 + x^2 = (x + 2)^2$ becomes $100 + x^2 = x^2 + 4x + 4$, which simplifies to
$4x = 96$, and therefore, $x = 24$.

$A = \frac{1}{2}ab = \frac{1}{2}(24)(10) = 120 \text{ cm}^2$

The area of a parallelogram can be found by multiplying the lengths of adjacent sides and the sine of the acute angle between them. Because the area of every triangle is one-half the area of some parallelogram, a similar rule can be used to find the area of a triangle. The area of a triangle is equal to one-half the product of any two adjacent sides and the sine of the angle included between them. In $\triangle ABC$, if a is the length of the side opposite $\angle A$, b is the side opposite $\angle B$, and c is the side opposite $\angle C$, the area of the triangle is

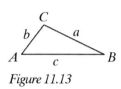

Figure 11.13

$$A = \tfrac{1}{2}ab\sin C = \tfrac{1}{2}ac\sin B = \tfrac{1}{2}bc\sin A$$

Example 4

In $\triangle ABC$, $BC = 14$ inches, $AB = 18$ inches, and the area of $\triangle ABC$ is 63 square inches. Find the measure of $\angle ABC$.

Solution: Start with the formula $A = \tfrac{1}{2}ac\sin B$ and substitute the known values to get $63 = \tfrac{1}{2}(14)(18)\sin B$ or $63 = 126\sin B$. Solve to find $\sin B = \dfrac{63}{126} = \dfrac{1}{2}$ and then use the inverse function to find the measure of the angle. $\sin^{-1}\left(\tfrac{1}{2}\right) = 30°$, so the measure of $\angle ABC$ is 30°.

If you don't know the height of the triangle or the measure of an angle, it is still possible to find the area of the triangle if the lengths of all three sides are known. The proof that **Heron's formula** actually gives you the area of the triangle involves either a rather complicated geometric proof or some information about trigonometry that you have not yet studied. Just know that it can be proved, by starting with $A = \tfrac{1}{2}ab\sin C$, and making a substitution that expresses $\sin C$ in terms of the sides of the triangle. After a good bit of simplifying, you arrive at Heron's formula:

$$A = \sqrt{s(s-a)(s-b)(s-c)} \cdot$$

The first step in finding the area of a triangle by Heron's formula is to calculate the semiperimeter. *Semi* means half, and the semiperimeter is literally half the perimeter. If the sides of the triangle are represented by

a, b, and c, then the semiperimeter is $s = \frac{a+b+c}{2}$. The area of the triangle

is $A = \sqrt{s(s-a)(s-b)(s-c)}$.

Example 5

Find the area of an isosceles triangle whose base measures 4 inches and whose legs each measure 8 inches.

Solution: The semiperimeter $s = \dfrac{a+b+c}{2} = \dfrac{4+8+8}{2} = 10$.

According to Heron's formula,

$A = \sqrt{s(s-a)(s-b)(s-c)} = \sqrt{10(10-4)(10-8)(10-8)}$. Simplify

to find $A = \sqrt{10(6)(2)(2)} = \sqrt{240} = 4\sqrt{15}$.

Lesson 11-2 Review

1. In $\triangle ABC$, $\overline{AC} \perp \overline{BC}$, $AC = 12$ cm, and $BC = 8$ cm. If $AB = 4\sqrt{13}$, find the length of $\overline{CD}$.

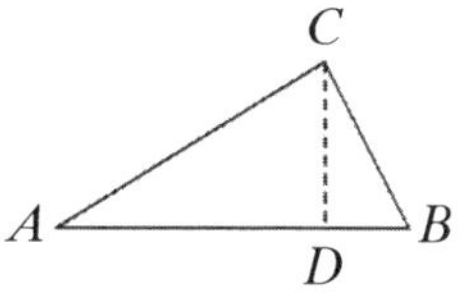

Figure 11.14

2. In $\triangle RST$, altitude $\overline{SV} \perp \overline{RT}$. SV is 3 inches longer than RT, and the area of $\triangle RST$ is 5 square inches. Find the length of $\overline{SV}$.

3. In right triangle $\triangle XYZ$, with $\overline{XY} \perp \overline{YZ}$, $\dfrac{XY}{YZ} = \dfrac{5}{12}$. If the area of $\triangle XYZ$ is 480 square meters, find the length of hypotenuse $\overline{XZ}$.

4. In $\triangle ABC$, $AB = 9$ inches, $BC = 13$ inches, and the measure of $\angle ABC$ is 20°. Find the area of $\triangle ABC$.

5. Find the area of an equilateral triangle with sides 16 cm long.

Lesson 11-3: Regular Polygons

No convenient formulas exist for polygons with more than four sides, unless those polygons are regular. **Regular polygons** have all sides and all angles congruent, and the formula depends on that regularity. It is derived by dividing the polygon into congruent triangles, calculating the area of one of those triangles, and multiplying by the number of triangles, that is, by the number of sides.

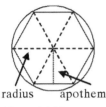

radius ⎯ apothem

Figure 11.15

Place the regular polygon in a circle so that each vertex lies on the circle. Drawing all the radii divides the polygon into congruent isosceles triangles. The altitude of one of these isosceles triangles is called the **apothem** of the polygon. The area of one of these triangles is one-half the product of its base and height, or one-half the product of a side and the apothem of the polygon. Multiply by the number of triangles—in this example, six—and the area of the polygon is $6 \cdot \frac{1}{2} \cdot s \cdot a$, where s is the length of a side and a is the length of the apothem. Rearranging this product to $\frac{1}{2}a \cdot 6s$, and remembering that $6s$ is the perimeter of the hexagon, you get the formula for the area of a regular polygon. If a is the length of the apothem and p is the perimeter, $A = \frac{1}{2}ap$.

Example 1

Find the area of a regular octagon with sides 5 cm long and an apothem of 7 cm.

Solution: The octagon has eight sides of 5 cm each, so its perimeter is 40 cm. The apothem is 7 cm, so the area of the octagon is $A = \frac{1}{2}(7)(40) = 140 \text{ cm}^2$.

Figure 11.16

Example 2

A regular pentagon has an area of 90 square centimeters. If the apothem measures 4 cm, find the length of one side of the pentagon.

Solution: $A = \frac{1}{2}ap$ becomes $90 = \frac{1}{2}(4)p$ when the known values are substituted. Solving for p gives $p = 45$ cm. The pentagon has 5 congruent sides, each of them

$$\frac{45}{5} = 9 \text{ cm.}$$

Figure 11.17

> ## Lesson 11-3 Review

1. Find the area of a regular hexagon with sides 12 cm long. [Hint: When the hexagon is divided into triangles, what kind of special triangles are formed?]

2. A regular decagon has an area of 400 square centimeters. If the side of the decagon measures 8 cm, find the length of the apothem.

Lesson 11-4: Circumference and Area of Circles

For circles, the term circumference designates the same type of measurement that is called perimeter for polygons. The **circumference** of a circle measures the distance around the circle. Ancient mathematicians discovered that the ratio of the circumference of a circle to its diameter was always a number a little greater than 3. Many attempts were made over the centuries to determine the exact value of that ratio, but the Greek letter π has become the accepted symbol for the constant. The circumference of a circle is given by the formula $C = \pi d = 2\pi r$, where d is the diameter of the circle and r is its radius. Traditionally, an approximate value of 3.14 or of $\frac{22}{7}$ is used for π, but those are approximate values. For an exact value, leave your answer in terms of π. The exact value of the circumference of a circle of diameter 7 inches is 7π inches. The approximate values of that circumference might be $7(3.14) = 21.98$ or $7\left(\frac{22}{7}\right) = 22$ inches.

The formula for the area of a circle can be determined by looking at rearrangements of pieces of the circle. Divide the circle into congruent wedges, called sectors, and arrange them with points alternating up and down as shown. The result is a figure that begins to resemble a parallelogram. (Cutting the circle into more, smaller sectors makes the figure look more similar to a parallelogram.) The top and bottom sides of this figure add up to the circumference of the circle, so its base is $\frac{1}{2}C = \pi r$. Its height is the radius of the circle. The **area of the circle** is the area of this near-parallelogram, which is base times height, or $A = \frac{1}{2}C \cdot r = \frac{1}{2}(\pi d)r = \frac{1}{2}(2\pi r)r = \pi r^2$.

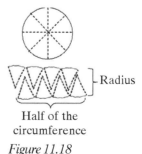

Half of the circumference

Figure 11.18

Example 1

Find the circumference of a circle whose area is 64π cm^2.

Solution: A $= \pi r^2$, so if the area is 64π cm^2, $r^2 = 64$ and $r = 8$ cm. If the radius is 8 cm, the diameter is 16 cm, and so the circumference is 16π cm.

Example 2

The circles shown are concentric (that is, both have center O). The area of the shaded region is 24π in^2 and $PQ = 2$ inches. Find OP.

Solution: The area of the shaded region is equal to the area of the larger circle minus the area of the smaller circle. Let $OP = x$ and $OQ = x + 2$. The area of the smaller circle is πx^2, and the area of the larger circle is $\pi(x + 2)^2$. Then $\pi(x + 2)^2 - \pi x^2 = 24\pi$ in^2. Divide through by π and the equation becomes $(x + 2)^2 - x^2 = 24$. Simplify and the left side becomes $x^2 + 4x + 4 - x^2 = 24$ or $4x + 4 = 24$. Solving gives $4x = 20 \Rightarrow x = 5$, so $OP = 5$ inches.

Figure 11.19

Lesson 11-4 Review

1. Find the diameter of a circle whose area (in square centimeters) is equal to its circumference (in centimeters).

2. A circle is inscribed in a square whose diagonal is 14 centimeters. Find the area of the shaded region.

Figure 11.20

Lesson 11-5: Arcs and Sectors

An **arc** is a portion of the circle intercepted by a central angle. The measure of an arc, in degrees, is equal to the measure of the central angle that intercepts it, and does not depend on the length of the radius. The length of an arc, however, is a portion of the circumference, and is dependent on both the measure of the central angle and the length of the radius.

The measure of $\overset{\frown}{AB}$ is equal to the measure of $\overset{\frown}{CD}$. If m$\angle COD = 70°$,

m$\overset{\frown}{AB} = 70°$ and m$\overset{\frown}{CD} = 70°$. Because radius $OD > OB$,

$\overset{\frown}{CD}$ is longer, in units such as centimeters or inches, than

$\overset{\frown}{AB}$. The actual length of an arc is a fraction of the circumference of the circle, and the fraction is simply the measure of the central angle over 360, the total degrees in the circle. If $OB = 3$ cm and m$\angle COD = 70°$, the length of

Figure 11.21

$\overset{\frown}{AB}$ is $\frac{\text{angle}}{360} \cdot 2\pi r = \frac{70}{360} \cdot 2\pi \cdot 3 = \frac{7\pi}{6} \approx 3.67$ cm. If $OD = 5$ cm, the length of

$\overset{\frown}{CD}$ will be $\frac{70}{360} \cdot 2\pi \cdot 5 = \frac{35\pi}{18} \approx 6.11$ cm.

Example 1

A central angle of 36° intercepts an arc that is 4 inches long. Find the radius of the circle.

Solution: The length of the arc is found by $\frac{36}{360} \cdot 2\pi r$.

The length of the arc is 4 inches, so solve $\frac{36}{360} \cdot 2\pi r = 4$ to find r.

$$\frac{\pi r}{5} = 4 \Rightarrow \pi r = 20 \Rightarrow r = \frac{20}{\pi} \approx 6.37 \text{ inches}$$

Example 2

m$\angle COD = 60°$. If the length of $\overset{\frown}{CD}$ is 8 centimeters and the length of $\overset{\frown}{AB}$ is 5 centimeters, find BD.

Solution: The length of $\overset{\frown}{CD}$ is $\frac{60}{360} \cdot 2\pi \cdot OD = 8$, so

Figure 11.22

$\frac{\pi}{3} OD = 8$ and $OD = \frac{24}{\pi}$. The length of $\overset{\frown}{AB}$ is $\frac{60}{360} \cdot 2\pi \cdot OB = 5$, so

$\frac{\pi}{3} OB = 5$ and $OB = \frac{15}{\pi}$. $BD = OD - OB = \frac{24}{\pi} - \frac{15}{\pi} = \frac{9}{\pi} \approx 2.86$ cm.

A *sector* of a circle is a portion of the area cut off by a central angle. The area of a sector can be calculated as a fraction of the area of the circle, just as the length of an arc is a fraction of the circumference. The same fraction is used: the measure of the central angle divided by 360 for the total degrees in the circle.

Example 3

Find the area of the sector of a circle cut off by a central angle of 120° if the diameter of the circle is 12 cm.

Solution: The area of the sector is equal to $\dfrac{120}{360} \cdot \pi r^2$. Because the diameter is 12 cm, the radius is 6 cm. The area of the sector is:

$$\frac{120}{360} \cdot \pi \cdot 6^2 = \frac{36\pi}{3} = 12\pi \text{ cm}^2 .$$

Example 4

In circle O, $\angle AOB$ is a right angle, and the length of $\overset{\frown}{AB}$ is 8π. Find the area of sector *AOB*.

Solution: $\angle AOB$ is a right angle and the length of $\overset{\frown}{AB}$ is 8π, so $\dfrac{90}{360} \cdot 2\pi r = \dfrac{\pi r}{2} = 8\pi$; therefore $r = 16$. The area of the sector is:

Figure 11.23

$$\frac{90}{360} \cdot \pi r^2 = \frac{1}{4} \cdot \pi \cdot 16^2 = 64\pi .$$

Lesson 11-5 Review

1. $\angle AOB$ is a central angle and intercepts an arc that is 20π cm long. If the radius of the circle is 50 cm, find the measure of $\angle AOB$.

2. Inscribed angle $\angle ABC = 40°$. If the length of $\overset{\frown}{AC}$ is 16π cm, find the radius of the circle.

3. Find the diameter of a circle if a 40° sector has an area of 49π square units.

4. In circle O, $\angle AOB$ measures 20° and $OA = 3$ cm. In circle P, $\angle CPD$ measures 5°, and the area of the sector CPD is equal to the area of sector AOB. Find the radius PC.

Answer Key

Lesson 11-1

1.
$$3x(3+x) = 84$$
$$9x + 3x^2 = 84$$
$$3x^2 + 9x - 84 = 0$$
$$3(x^2 + 3x - 28) = 0$$
$$3(x+7)(x-4) = 0$$
$$x + 7 = 0 \Rightarrow x = -7$$
$$x - 4 = 0 \Rightarrow x = 4$$

The length of the rectangle is $3(4) = 12$ feet and the width is $3 + 4 = 7$ feet.

2. In any square with side s and diagonal d,

$s^2 + s^2 = d^2$ or $2s^2 = d^2$. Dividing by 2, $s^2 = \frac{1}{2}d^2$.
The area of the square is $A = s^2$,

so by substitution, $A = \frac{1}{2}d^2$.

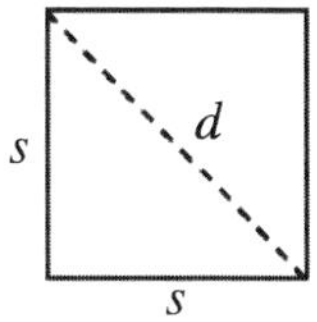

Figure 11.24

3. $A_{\square ABCD} = A_{\square WXYZ}$

$$bh = \frac{1}{2}d_1 d_2$$
$$AB \cdot h = \frac{1}{2} \cdot WY \cdot XZ$$
$$AB \cdot h = \frac{1}{2} \cdot AB \cdot AD$$
$$h = \frac{1}{2}AD = \frac{1}{2}(14) = 7$$

4.
$$5^2 + x^2 = 13^2$$
$$25 + x^2 = 169$$
$$x^2 = 144$$
$$x = 12$$
$$BD = 2(12) = 24$$

Area $= \frac{1}{2}(10)(24) = 120\,\text{cm}^2$

5. $A = \frac{1}{2}(b_1 + b_2)h$

   $35 = \frac{1}{2}(3 + h)h$

   $70 = 3h + h^2$

   $h^2 + 3h - 70 = 0$

   $(h + 10)(h - 7) = 0$

   $h + 10 = 0 \Rightarrow = -10$

   $h - 7 = 0 \Rightarrow h = 7$ meters

6. $A_{\square ABCD} = \frac{1}{2} \cdot 8 \cdot 14 = 56$.

   The area of the shaded region $= 56 - 4(2^2) = 56 - 16 = 40 \, \text{cm}^2$.

Lesson 11-2

1. $A_{\triangle ABC} = \frac{1}{2}(8)(12) = 48 \, \text{cm}^2$

   $A_{\triangle ABC} = \frac{1}{2}bh = \frac{1}{2}(4\sqrt{13})h = 2\sqrt{13}h$

   $2\sqrt{13}h = 48$

   $h = \dfrac{48}{2\sqrt{13}} = \dfrac{48\sqrt{13}}{2(13)} = \dfrac{24\sqrt{13}}{13} \, \text{cm}$

2. $A = \frac{1}{2}bh$

   $5 = \frac{1}{2}x(x + 3)$

   $10 = x^2 + 3x$

   $x^2 + 3x - 10 = 0$

   $(x + 5)(x - 2) = 0$

   $x + 5 = 0 \Rightarrow x = -5$

   $x - 2 = 0 \Rightarrow x = 2$

   $SV = x + 3 = 2 + 3 = 5$ inches

3. $\dfrac{XY}{YZ} = \dfrac{5}{12} \Rightarrow 5YZ = 12XY \Rightarrow YZ = \dfrac{12}{5}XY$

$A = \dfrac{1}{2}XY \cdot YZ$

$480 = \dfrac{1}{2}XY \cdot \dfrac{12}{5}XY$

$480 = \dfrac{6}{5}(XY)^2$

$\dfrac{5}{6}(480) = 400 = (XY)^2$

$XY = 20$

$\dfrac{XY}{YZ} = \dfrac{5}{12} \Rightarrow \dfrac{20}{YZ} = \dfrac{5}{12} \Rightarrow 5YZ = 240 \Rightarrow YZ = 48$

$(XY)^2 + (YZ)^2 = (XZ)^2$

$20^2 + 48^2 = (XZ)^2$

$400 + 2304 = 2704 = (XZ)^2$

$XZ = 52$ meters

4. $A = \frac{1}{2}AB \cdot BC\sin B$

$A = \frac{1}{2} \cdot 9 \cdot 13 \cdot \sin 20°$

$A = 20.01$ in^2

5. $s = \dfrac{16 + 16 + 16}{2} = 24$, and $s - a = s - b = s - c = 24 - 16 = 8$

$A = \sqrt{s(s-a)(s-b)(s-c)}$

$A = \sqrt{24(8)^3} = \sqrt{3(8)^4} = 8^2\sqrt{3} = 64\sqrt{3}$ cm

Lesson 11-3

1. Each triangle is equilateral. $a = \dfrac{12}{2}\sqrt{3} = 6\sqrt{3}$. The perimeter is $6(12) = 72$.

The area is $\frac{1}{2}(72)\left(6\sqrt{3}\right) = 216\sqrt{3}$ cm^2.

2. $p = 10(8) = 80$

$A = \frac{1}{2}ap$

$400 = \frac{1}{2}a \cdot 80$

$5 = \frac{1}{2}a$

$a = 10$

Lesson 11-4

1. $\pi r^2 = 2\pi r$

$r^2 - 2r = 0$

$r(r-2) = 0$

$r = 0 \text{ or } r = 2$

The radius is 2 cm, so the diameter is 4 cm.

2. $s = \dfrac{14}{\sqrt{2}} = 7\sqrt{2}$ cm, $A_\square = \left(7\sqrt{2}\right)^2 = 98 \text{ cm}^2$, $r = \dfrac{7\sqrt{2}}{2}$, and $A_\circ = \dfrac{49\pi}{2} \text{ cm}^2$.

The area of the shaded region is $98 - \dfrac{49\pi}{2} = \dfrac{196 - 49\pi}{2} \approx 21.03 \text{ cm}^2$.

Lesson 11-5

1.   The measure of the arc is equal to $\dfrac{m\angle AOB}{360} \cdot 2\pi r$.

$20\pi = \dfrac{x}{360} \cdot 2\pi \cdot 50$

$20\pi = \dfrac{5x}{18} \cdot \pi$

$20 = \dfrac{5}{18}x$

$20 \cdot \dfrac{18}{5} = x$

$x = 72°$

$m\angle AOB = 72°$

2. Because inscribed angle $\angle ABC = 40°$, $\overset{\frown}{AC}$ measures 80°, and so the central angle $\angle AOC$ measures 80°. The length of $\overset{\frown}{AC}$ is $\dfrac{80}{360}$ times the circumference of the circle.

$$\frac{80}{360}(2\pi r) = 16\pi$$

$$\frac{2}{9}(2\pi r) = 16\pi$$

$$\frac{4\pi r}{9} = \frac{16\pi}{1}$$

$$4\pi r = 144\pi$$

$$r = \frac{144\pi}{4\pi} = 36$$

The radius of the circle is 36 cm.

3. The area of a 40° sector is equal to $\dfrac{40}{360}\pi r^2$ and that is equal to 49π square units.

$$\frac{40}{360}\pi r^2 = 49\pi$$

$$\frac{\pi r^2}{9} = \frac{49\pi}{1}$$

$$\pi r^2 = 441\pi$$

$$r^2 = 441$$

$$r = 21$$

The diameter of the circle is twice the radius, or 42 units.

4. The area of the sector CPD is $\dfrac{20}{360}\pi r^2 = \dfrac{20}{360}\pi \cdot 3^2 = \dfrac{\pi}{2}$. The area of sector AOB is $\dfrac{5}{360}\pi r^2 = \dfrac{\pi r^2}{72}$. Because the two sectors have equal areas, $\dfrac{\pi}{2} = \dfrac{\pi r^2}{72}$.
Then $72\pi = 2\pi r^2$, so $36 = r^2$ and $6 = r$.
PC = 6 cm.

Geometric Solids

Lesson 12-1: Polyhedra

Most of your study of geometry is plane geometry, a look at figures that exist in the plane, such as polygons and circles. Solid geometry is concerned with three-dimensional figures, objects that exist in space. What we commonly refer to as a solid is a polyhedron, an object in space made up of polygons. The polygons are the faces of the polyhedron, the line segments where the polygons meet are the edges, and the points or corners where edges meet are vertices.

Just as polygons are named according to the number of their sides, polyhedra take their names from the number of faces they have. A tetrahedron has four faces, a hexahedron has six, and an octahedron has eight faces.

A polyhedron is regular if its faces are congruent, regular polygons. That definition sounds simple, but there are, in fact, only five convex regular polyhedra. The regular tetrahedron is formed from four equilateral triangles. The regular hexahedron—or, by its more common name, the cube—has six squares as faces. The regular octahedron has eight faces, all equilateral triangles. The dodecahedron is formed from pentagons, 12 of them in total. Twenty equilateral triangles form a regular icosahedron.

Figure 12.1

If you examine some polyhedra, whether regular or not, and start to count faces, edges, and vertices, you may begin to notice a pattern.

	Faces	Vertices	Edges
Tetrahedron	4	4	6
Hexahedron	6	8	12
Octahedron	8	12	18
Dodecahedron	12	20	30

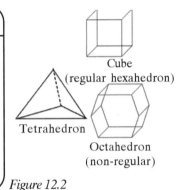

Cube
(regular hexahedron)

Tetrahedron

Octahedron
(non-regular)

Figure 12.2

The number of faces plus the number of vertices, minus the number of edges, always equals two. This formula, attributed to Euler, can be stated as $F + V - E = 2$.

Example 1

A polyhedron has 30 edges and 12 vertices. Name the polyhedron.

Solution: $F + V - E = 2$, so $F + 12 - 30 = 2$, which means $F = 20$. A polyhedron with 20 faces is an icosahedron.

Lesson 12-1 Review

1. The Fullerene molecule, nicknamed Bucky Ball and named for Buckminster Fuller, is a polyhedron with 32 faces. Twelve of these are pentagons, and 20 are hexagons. If the Bucky Ball has 90 edges, how many vertices does it have?

Lesson 12-2: Prisms

A **prism** is a polyhedron with two parallel faces, called bases, and parallelograms for its other faces. If the non-parallel faces are rectangles, and are perpendicular to the bases, the prism is a right prism. A prism takes its name from the polygons that form its bases. If the bases are pentagons, the figure is a pentagonal prism. Triangular prisms have triangles as their bases.

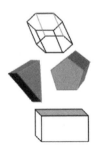

Figure 12.3

The **surface area of a polyhedron** is the total of the areas of the polygons that form its faces. For prisms, this means the area of two bases and several parallelograms.

Example 1

A right triangular prism has bases that are right triangles with legs 3 cm and 4 cm long. Find the surface area of the prism if it is 6 cm high.

Solution: First, use the Pythagorean Theorem to find the length of the hypotenuse of the right triangle $(3^2 + 4^2 = c^2 \Rightarrow c = 5)$. The area of each base is $\frac{1}{2}(3)(4) = 6$ cm^2. The other faces are rectangles.

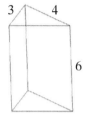

Figure 12.4

One is 3 by 6, with an area of 18 cm^2. Another face is 4 by 6, with an area of 24 cm^2. The third face is 5 by 6, with an area of 30 cm^2. The surface area of the prism is therefore $2(6) + 18 + 24 + 30 = 84$ cm^2.

In general, the surface area of a right prism is $S = 2(A_{base}) + h(P_{base})$, where A_{base} designates the area of the base and P_{base} is the perimeter of the base. If the base is a regular polygon, the formula becomes $S = 2(\frac{1}{2}aP_{base}) + h(P_{base}) = P_{base}(a + h)$, where a is the apothem of the base and h is the height of the prism.

Example 2

In the right pentagonal prism, the apothem of the regular pentagon is 7 cm, the sides of the pentagon are 19 cm each, and the height of the prism is 15 cm. Find the surface area of the prism.

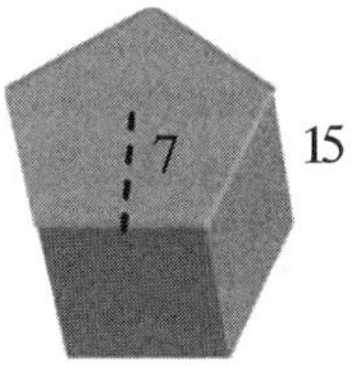

Solution: The surface area is equal to the perimeter of the base, which is $5(19) = 95$ cm, times the sum of the apothem of the polygon and height of the prism. $S = P_{base}(a + h)$ becomes $S = 95(7 + 15) = 95(22) = 2{,}090$. The surface area of the pentagonal prism is 2,090 cm^2.

Figure 12.5

If area can be visualized as a count of the number of one-unit square tiles that are needed to cover a surface, then volume can be thought of as the number of cubes, one unit on an edge, that are needed to fill the interior of the polyhedron. Such an image becomes unwieldy, however, for any polyhedra other than rectangular prisms.

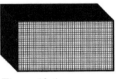

Figure 12.6

Fortunately, the rectangular prism gives rise to a formula for volume that is valid for any prism. If you were packing a rectangular prism with cubes, you would first make a layer on the bottom, then stack additional layers on top, until the box was filled. The number of cubes necessary to make one layer will be equal to the area of the base. The number of layers will be equal to the height.

Generalizing that rule, the volume of a prism is equal to the area of the base times the height ($V = Bh$). The rule for finding the area of the base will vary depending on the type of polygon that forms the base. If the polygon is regular, you can use the formula $A = \frac{1}{2}ap$.

Example 3

Find the volume of a rectangular prism 5.2 cm long, 8.1 cm wide, and 3.7 cm high.

Solution: The volume of the prism is the area of the base (5.2 cm × 8.1 cm) times the height (3.7 cm).

$V = 5.2 \times 8.1 \times 3.7 = 155.844$ cm^3

Example 4

A hexagonal prism has bases that are regular hexagons with sides 8 inches long. If the prism is 18 inches high, find its volume.

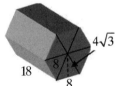

Figure 12.7

Solution: First, realize that the diagonals divide the hexagon into six equilateral triangles, and use 30–60–90 triangle relationships to determine that the apothem has a length of $4\sqrt{3}$. The base is a hexagon, so six sides of length 8 give it a perimeter of 48 inches. The area of the base is $A = \frac{1}{2}ap = \frac{1}{2}(4\sqrt{3})(48)$.

The height of the prism is 18, so the volume is

$V = \frac{1}{2}(4\sqrt{3})(48)(18) = 1728\sqrt{3} \approx 2992.98$.

The volume of the hexagonal prism is approximately 2,993 in³.

Lesson 12-2 Review

1. A right octagonal prism has bases that are regular octagons with sides 5 inches long and an apothem 6 inches long. Find the surface area of the prism if it is 12 inches high.

2. In Figure 12.8, the perimeter of the hexagon is 42 cm, and the height of the prism is 20 cm. Find the surface area of the prism.

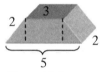

Figure 12.8

3. Find the surface area and the volume of a triangular prism 14 inches high, if its base is a triangle with sides 9, 13, and 20 inches long.

4. A trapezoidal prism has bases that are isosceles trapezoids with non-parallel sides 2 cm long, and parallel sides of 3 cm and 5 cm. If the parallel bases are 2 cm apart, find the volume of the prism.

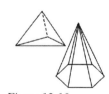

Figure 12.9

Lesson 12-3: Pyramids

Pyramids are formed from triangles around a polygonal base. The triangles meet at a single point. Most of us, hearing the word *pyramid*, have an image of the pyramids of Egypt or Central America. Their bases are square, or nearly square, and are surrounded by four triangles. The base of a pyramid need not be square, however, and the pyramid takes its name from the polygon that forms its base. A tetrahedron is a triangular pyramid. A pyramid that has a six-sided base is a hexagonal pyramid.

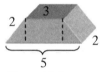

Figure 12.10

The height of a pyramid, *h*, is the length of a perpendicular to the base, dropped from the point where the triangular faces meet. The slant height, *l*, is the altitude of the triangles that form the other faces.

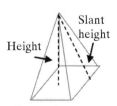

Figure 12.11

The **surface area of a pyramid** is the sum of the areas of its faces. If the base is a regular polygon, the surrounding triangular faces will be congruent. The surface area will be equal to the area of the regular n-gon plus the areas of the n triangles surrounding it. Therefore

$$S = \tfrac{1}{2}ap + n\left(\tfrac{1}{2}bl\right) = \tfrac{1}{2}ap + \tfrac{1}{2}pl = \tfrac{1}{2}p(a+l).$$

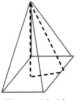

If the base is not regular, however, the triangular faces must be examined separately.

Figure 12.12

Example 1

Find the surface area of a square pyramid if the edge of the base is 6 inches and the height is 4 inches.

Solution: The base of the pyramid is a square, 6 inches on a side, so its area is 36 square inches. Because the square is regular, the four triangles will be congruent, and therefore, you can find the area of one and multiply *Figure 12.13* by four. To find the area of one of the triangles, you need the base, which is the 6-inch edge, and the height of the triangle, which is the slant height of the pyramid. That measurement is not given, so you need to think about how you can find it.

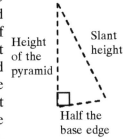

If you draw the line segments representing the height of the pyramid and the slant height, and connect them with a line segment on the plane of the base, you form a right triangle. That right triangle has legs that are the height of the pyramid and half of the side of the square that forms the base. Its hypotenuse is the slant height. The height of the pyramid is 4 inches and half of the base edge is half of 6, or 3 inches. The triangle is, therefore, a 3–4–5 right triangle, so the slant height is 5 inches.

Height of the pyramid

Slant height

Half the base edge

Figure 12.14

The area of one of the triangular faces is $A = \tfrac{1}{2}bh = \tfrac{1}{2}(6)(5) = 15$, so, multiplying by 4, the total area of the triangular faces is 60 square inches. The total surface area of the pyramid is 36 + 60 = 96 square inches.

Example 2

A right pyramid has a regular hexagon, 8 cm on each edge, as its base. If the surface area of the pyramid is $336\sqrt{3}$ cm², find the slant height of the pyramid.

Solution: $S = \frac{1}{2}ap + 6\left(\frac{1}{2} \cdot 8l\right)$. The surface area is $336\sqrt{3}$ and the perimeter is $6 \cdot 8 = 48$, so the equation becomes

$$336\sqrt{3} = \frac{1}{2}a \cdot 48 + 6\left(\frac{1}{2} \cdot 8l\right).$$

As noted before, the diagonals of the regular hexagon create equilateral triangles, and using the 30–60–90 triangle created by drawing the apothem, you can conclude that $a = 4\sqrt{3}$. So the equation becomes $336\sqrt{3} = \frac{1}{2} \cdot 4\sqrt{3} \cdot 48 + 6\left(\frac{1}{2} \cdot 8l\right)$, which simplifies to $336\sqrt{3} = 96\sqrt{3} + 24l$. Solving gives you $240\sqrt{3} = 24l$ and $l = 10\sqrt{3}$. The slant height of the pyramid is $10\sqrt{3}$ cm.

If you compare the volumes of a triangular prism and a triangular pyramid with the same bases and heights, it is clear that the volume of the pyramid is smaller—specifically, one-third the volume of a similarly sized prism. For pyramids, $V = \frac{1}{3}Bh$, where B is the area of the base polygon and h is the height of the pyramid.

Example 3

Find the volume of a square pyramid 24 meters on a side and 30 meters high.

Solution: The base is a square, so the area of the base is $24^2 = 576$ m², so $V = \frac{1}{3}(576)(30) = 5,760$ m³. The volume of the pyramid is 5,760 cubic meters.

Example 4

The base of a right pyramid is a regular pentagon with apothem $a = 4$ cm. The slant height of the pyramid is $l = 5$ cm. If the surface area of the pyramid is 135 cm², find its volume.

Solution: In order to find the volume, you must know the area of the base and the height. Neither is directly given. The surface area is $S = \frac{1}{2}p(a+l)$, so substituting the known values gives you $135 = \frac{1}{2}p(4+5)$ and solving shows that $p = 30$, which means that each side of the pentagon is 6 cm. You still need the height, so use the right triangle formed by the height, the apothem, and the slant height. The apothem is $a = 4$ cm, and the slant height is $l = 5$ cm, so by the Pythagorean relationship the height must be $h = 3$ cm. The volume of the pyramid is $V = \frac{1}{3}Bh$, and the area of the base is

$\frac{1}{2}ap$, so $V = \frac{1}{3}\left(\frac{1}{2}ap\right)h = \frac{1}{6}aph = \frac{1}{6}(4)(30)(3) = 60$ cm^3.

Lesson 12-3 Review

1. If the surface area of a triangular pyramid is $130\sqrt{3}$ cm^2, and the base is an equilateral triangle with an edge of 10 cm, find the slant height.

2. A pyramid has a square, 4 inches on a side, as its base. The slant height of the pyramid is 6 inches. Find the surface area of the pyramid.

3. Find the volume of a pyramid 3 meters high if the base is an octagon with an area of 24 square meters.

4. The volume of a hexagonal pyramid is 9,720 cm^3. If the side of the regular hexagon is $12\sqrt{3}$ cm, find the height of the pyramid.

Lesson 12-4: Cylinders and Cones

A **cylinder** is a three-dimensional object with two parallel circles as its bases. A **cone** has a single circle as its base, and the lateral surface comes to a point. Informally, you might think of a cylinder as a circular prism and a cone as a circular pyramid. Much of what you know about prisms and pyramids can translate to cylinders and circles.

Cylinder

Cone

Figure 12.15

The **surface area of a cylinder** is equal to the sum of the areas of the two circular bases plus the lateral area. If you think of the cylinder as, for example, a soup can, the lateral area is the area of the label wrapped around the can, which is a rectangle. Its height is the height of the cylinder, and its width is the circumference of the circle (because it must wrap all the way around the circle). The surface area, therefore, is the area of the two circles plus the area of the rectangle. The area of each circle is πr^2, the circumference is $2\pi r$, and the height is h, so the surface area of the cylinder can be given by the formula $S = 2(\pi r^2) + (2\pi r)h = 2\pi r(r + h)$.

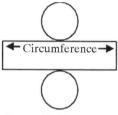

Figure 12.16

Example 1

Determine how much aluminum is necessary to make a cylindrical can 5 inches high and 3 inches in diameter.

Solution: The amount of aluminum needed is the surface area of the can. The height is 5 inches and the radius is half the diameter, so $r = 1.5$ inches.
$S = 2\pi r(r + h) = 2\pi(1.5)(1.5 + 5) = 3\pi(6.5) = 19.5\pi$.
Manufacturing the can requires 19.5π square inches, or approximately 61.26 square inches, of aluminum.

The surface area of a cone is more difficult to visualize. The base is a circle with an area equal to πr^2, but the lateral area is harder to see. Imagine that you took a paper cone, the kind that's sometimes used to hold a snow cone, and you split the cone from the open edge to the point. If you flatten the paper out, you have part of a circle, a sector. This circle will be larger than the circle that forms the base of the cone. The radius of the large circle is the cut you made up the side of the cone, that is, the slant height. To determine what part of the large circle you have, compare the length of the arc of your sector to the circumference of the large circle. When this sector was rolled into a cone, the arc was forming the circumference of the circular base, so its length is $2\pi r$. The circumference of the large circle is $2\pi l$, so the fraction you have is $\frac{2\pi r}{2\pi l} = \frac{r}{l}$.

Figure 12.17

The area of your sector will be that fraction of the area of the large circle, so the lateral area of the cone is $\frac{r}{l} \cdot \pi l^2 = \pi r l$.

The total surface area of the cone is the area of the circular base plus this lateral area. That gives you the formula $S = \pi r^2 + \pi r l = \pi r(r + l)$. To find the surface area of a cone, you need to know its radius and its slant height. Remember that the height of the cone, the radius of the circular base, and the slant height will form a right triangle, and therefore, by the Pythagorean Theorem, $r^2 + h^2 = l^2$.

Example 2

Find the surface area of a cone with a diameter of 8 cm and a height of 12 cm.

Solution: To find the surface area of a cone, you need the radius and the slant height. The diameter is 8 cm, so the radius is 4 cm; you can use the Pythagorean relationship to find the slant height.

$r^2 + h^2 = l^2$ so $4^2 + 12^2 = 16 + 144 = 160 = l^2$. This means

$l = \sqrt{160} = 4\sqrt{10}$. The surface area is

$S = \pi r(r + l) = \pi \cdot 4(4 + 4\sqrt{10})$, which simplifies to

$S = 16\pi + 16\pi\sqrt{10} = 16\pi(1 + \sqrt{10}) \approx 209.22$ cm^2 .

The volume of a cylinder can be found just the way the volume of a prism is found: the area of the base times the height. In the case of the cylinder, the base is a circle and so the formula for the volume can be given as $V = \pi r^2 h$.

Example 3

Find the volume of a cylindrical can 7 inches high and 4 inches in diameter.

Solution: The diameter is 4 inches, so the radius is 2 inches. Substituting in the formula $V = \pi r^2 h$, you have $V = \pi \cdot 2^2 \cdot 7 = 28\pi$. The volume of the can is 28π cubic inches, or approximately 87.96 cubic inches.

To find the volume of a cone, you can apply what you learned about pyramids. The volume of the cone is one-third the volume of a cylinder with the same radius and height, so $V = \frac{1}{3}\pi r^2 h$.

Example 4

Two cones are constructed, both with a radius of 4 inches. The cones have heights of 6 inches and 9 inches. How much more will the taller cone hold?

Solution: The volume of the shorter cone is

$V = \frac{1}{3}\pi r^2 h = \frac{1}{3}\pi (4)^2 (6) = 32\pi$ in^3, and the volume of the taller one

is $V = \frac{1}{3}\pi r^2 h = \frac{1}{3}\pi (4)^2 (9) = 48\pi$ in^3. The difference in the volume

is $48\pi - 32\pi = 16\pi$. The taller cone will hold an additional 16π in^3, or approximately 50.27 in^3.

Lesson 12-4 Review

1. If a cylindrical container 12 inches high and 10 inches in diameter is formed from a 24 inch square sheet of cardboard, how much cardboard is wasted?

2. Which requires more material: a cone 20 cm high and 12 cm in diameter, or a cone 12 cm high and 20 cm in diameter?

3. A cylinder 7 inches high and 6 inches in diameter is filled to capacity, and the contents are to be poured into a cylinder 4 inches in diameter. How tall must the thinner cylinder be in order to match the volume of the first one?

4. The contents of a cylinder are to be transferred to a cone of the same height. If the radius of the cylinder is 3 cm, and its height is 5 cm, what must be the radius of the cone?

Lesson 12-5: Spheres

The **sphere**—the figure that might be described as a ball—is the three-dimensional object you would produce if you rotated a circle around its diameter. If you could grab the ends of the diameter and spin the circle

Figure 12.18

quickly, it might appear that there were many identical circles, at different angles, blurring together into the sphere. The radius and diameter of that original circle are the radius and diameter of the sphere.

Deriving the formulas for the surface area and volume of a sphere requires calculus, a branch of mathematics you have not yet encountered. With a little imagination, however, you can see that the formulas do make sense.

Imagine that you put the sphere inside a cylindrical can, just big enough to hold it. Both the diameter and the height of the cylinder are equal to the diameter of the sphere. Imagine that the sphere is an orange, and that you carefully make cuts in the top and bottom of the orange so that you can peel back the skin. If you pull back the sections of peel at the top and bottom so that they stand up, the orange peel will almost match the wall, or lateral surface, of the cylinder. It will be a little taller than the cylinder in some places, but there will be spaces between the sections. The gaps and the extra points poking up will cancel one another out. The wall of the cylinder—just the lateral area, without the circles—is $2\pi rh = 2\pi r(2r)$. The formula for the **surface area of a sphere** is $S = 4\pi r^2$, where r is the radius of the sphere.

Figure 12.19

Example 1

Selena is blowing up a spherical balloon. The first time she pauses for breath, the radius of the balloon is 5 cm. When she is done, the radius is 8 cm. How much greater is the surface area of the balloon when Selena is done than when she paused for breath?

Solution: When the radius is 5 cm, the surface area is $S = 4\pi r^2 = 4\pi \cdot 5^2 = 100\pi$ cm². When the radius grows to 8 cm, the surface area increases to $S = 4\pi \cdot 8^2 = 256\pi$ cm². Subtracting $256\pi - 100\pi = 156\pi$, so the final surface area is 156π, or 490.09 cm², greater than the surface when Selena paused.

If you put the sphere back into that cylindrical container, you can see that the **volume of the sphere** is less than the volume of the cylinder. The cylinder has a volume of $V = \pi r^2 h$, but, because the height of the cylinder is equal

Figure 12.20

to the diameter of the sphere, that becomes $V = \pi r^2 h = \pi r^2 (2r) = 2\pi r^3$. The volume of the sphere is smaller than $2\pi r^3$; the actual formula for the volume of a sphere is $V = \frac{4}{3}\pi r^3$.

Example 2

If a volleyball, 66 cm in diameter, is packed into a box that is a cube, how much empty space is in the box?

Solution: The diameter of the ball is 66 cm, so its radius is 33 cm. The volume of the volleyball is

$$V = \frac{4}{3}\pi r^3 = \frac{4}{3}\pi(33)^3 = \frac{4}{3}\pi(35{,}937) = 47{,}916\pi \text{ cm}^3.$$

The dimensions of the box must be, at a minimum, the diameter of the ball. A cube, 66 cm on each edge, will have a volume of 287,496 cm³. The empty space will be the volume of the box minus the volume of the ball, so $287{,}496 - 47{,}916\pi$ cm³, which is approximately 136, 963.45 cm³.

Lesson 12-5 Review

1. Find the volume of a sphere whose surface area is 324π cm².

2. Find the radius of a sphere if the numerical value of its volume and the numerical value of its surface area are equal.

3. Tennis balls are packed three in a cylindrical can, so that the balls touch one another and touch the sides of the can. If the diameter of a tennis ball is approximately 2.5 inches, how much empty space is inside the can?

4. Which is greater: the surface area of one sphere with a radius of 10 cm or the total surface area of two spheres, each with a radius of 5 cm?

Answer Key
Lesson 12-1

1. $F + V - E = 2$

   $32 + V - 90 = 2$

   $V = 2 - 32 + 90 = 60$

   The polyhedron has 60 vertices.

Lesson 12-2

1. $720 \, in^2$

2. $S \approx 1,094.61 \, cm^2$

3. $S \approx 677.8 \, in^2, V \approx 628.6 \, in^3$

4. $V = 4\sqrt{3} \cdot 2 = 8\sqrt{3} \, cm^3$

Lesson 12-3

1. $l = 7\sqrt{3} \, cm$

2. $S = 4^2 + 4\left(\frac{1}{2}\right)(4)(6) = 16 + 48 = 64 \, in^2$

3. $V = \frac{1}{3} \cdot 24 \cdot 3 = 24 \, m^3$

4. $15\sqrt{3}$

Lesson 12-4

1. The area of the cardboard is $24^2 = 576$ square inches.
   The surface area of the cylinder is
   $2\pi r^2 + 2\pi rh = 2\pi \cdot 5^2 + 2\pi \cdot 5 \cdot 12 = 50\pi + 120\pi = 170\pi.$.
   The wasted cardboard is $576 - 170\pi \approx 41.93$ square inches.

2. The first cone has a surface area of $S_1 = 36\pi + 12\pi\sqrt{109} \approx 506.69 \, cm^2$ and the second cone has a surface area of $S_2 = 100\pi + 20\pi\sqrt{61} \approx 804.89 \, cm^2$. The cone that is 12 cm high and 20 cm in diameter requires slightly less than 300 cm^2 more material.

3. The volume of the first cylinder is $\pi r^2 h = \pi \cdot 3^2 \cdot 7 = 63\pi$. The thinner cylinder has a volume of $\pi r^2 h = \pi \cdot 2^2 \cdot h = 4\pi h$, so $4\pi h = 63\pi$, and $h = \dfrac{63}{4} = 15.75$.
   The second cylinder must be 15.75 inches tall.

4. The volume of the cylinder is $\pi r^2 h = \pi \cdot 3^2 \cdot 5 = 45\pi$.
   The volume of the cone is $\dfrac{1}{3}\pi r^2 h = \dfrac{1}{3}\pi r^2 \cdot 5 = \dfrac{5}{3}\pi r^2$.
   In order for the cone to hold the contents of the cylinder, $\dfrac{5}{3}\pi r^2 = 45\pi$.
   Solving, you find $r^2 = 45 \cdot \dfrac{3}{5} = 27$ and $r = \sqrt{27} = \sqrt{9 \cdot 3} = 3\sqrt{3}$.
   The radius must be $3\sqrt{3}$ cm or approximately 5.20 cm.

Lesson 12-5

1. $S = 4\pi r^2$ so $324\pi = 4\pi r^2$ means that $r^2 = 81$ and $r = 9$.

   Then $V = \frac{4}{3}\pi r^3$ becomes $V = \frac{4}{3}\pi \cdot 9^3 = 972\pi$.

   The volume is 972π cm^3.

2. The numerical values—but not the units—of the volume and surface area are

   the same, so $\frac{4}{3}\pi r^3 = 4\pi r^2$. Then $4r^3 = 12r^2$, and $r^3 = 3r^2$. This equation has two

   possible solutions: $r = 0$ or $r = 3$. A radius of zero would not make sense. The radius is 3 units.

3. The can has a diameter equal to that of the tennis ball and a height equal to the total of the diameters of the three balls stacked in the can. So its diameter is 2.5 inches, its radius is 1.25 inches, and its height is 7.5 inches. The volume of the cylindrical can is $\pi r^2 h = \pi(1.25)^2(7.5) \approx 11.72\pi$. The volume of one tennis

   ball is $V = \frac{4}{3}\pi(1.25)^3$ or $\approx 2.604\pi$. The space taken up by the three tennis balls

   will be $\approx 3(2.604\pi) \approx 7.81\pi$. The empty space is $11.72\pi - 7.81\pi = 3.91\pi$ cubic inches, or approximately 12.28 cubic inches.

4. The surface area of the single sphere with radius 10 cm is
   $S_{large} = 4\pi r^2 = 4\pi \cdot 10^2 = 400\pi$ cm^2. The combined area of two smaller spheres is $2S_{small} = 2(4\pi r^2) = 2(4\pi \cdot 5^2) = 200\pi$ cm^2. The one larger sphere has twice the surface area of the two small spheres combined.

Index

About the Author

CAROLYN WHEATER teaches middle school and upper school mathematics at the Nightingale-Bamford School in New York City. Educated at Marymount Manhattan College and the University of Massachusetts, Amherst, she has taught math and computer technology for 30 years to students from preschool through college.

HOMEWORK HELPERS™

The Essential Help You Need When Your Textbooks Just Aren't Making the Grade!

Homework Helpers: Earth Science
Phil Medina
6 x 9, paper, 352 pp
ISBN 1-56414-767-3, U.S. $14.99 (Can. $19.95)
Illustrated

Homework Helpers: Biology
Matthew Distefano
6 x 9, paper, 352 pp
ISBN 1-56414-720-7, U.S. $14.99 (Can. $19.95)
Illustrated

Homework Helpers: Physics
Greg Curran
6 x 9, paper, 352 pp
ISBN 1-56414-768-1, U.S. $13.99 (Can. $19.95)
Illustrated

Homework Helpers: Chemistry
Greg Curran
6 x 9, paper, 352 pp
ISBN 1-56414-721-5, U.S. $14.99 (Can. $19.95)
Illustrated

Homework Helpers: Basic Math and Pre-Algebra
Denise Szecsei
6 x 9, paper, 224 pp
ISBN 1-56414-873-4, U.S. $13.99 (Can. $18.50)
Illustrated

Homework Helpers: Algebra
Denise Szecsei
6 x 9, paper, 224 pp
ISBN 1-56414-874-2, U.S. $13.99 (Can. $18.50)
Illustrated

Homework Helpers: English Language and Composition
Maureen Lindner
6 x 9, paper, 352 pp
ISBN 1-56414-812-2, U.S. $13.99 (Can. $18.95)
Illustrated

Great preparation for the SAT II and AP Courses

CAREER
PRES